First Published – 2008

Reprinted – 2026

ISBN: 978-81-8356-337-6

Introductory Business Mathematics

Published by:

DISCOVERY PUBLISHING HOUSE
4383/4B, Ansari Road, Darya Ganj
New Delhi-110 002 (India)
Phone: +91-11-23279245; 23253475; 43596065
Mobile: +91 9811179893 / +91 9871656464
E-mail: discoverybooksindia@gmail.com
orderdphbooks@gmail.com
namitwasan9@gmail.com
web: www.discoverypublishinggroup.com

Printed at:
Infinity Imaging Systems
Delhi

Preface

This book "Introductory Business Mathematics" has been written to provide a comprehensive and updated foundation course in Mathematics and its varied applications in management, commerce and economics.

This book has been written from studies point of view, so that they can easily understand various mathematical concepts, techniques and tool needed for their course. Each new concept and technique is properly supported by suitable solved examples. In solving the questions cave has been taken to explain each step so that students can follow the subject matter themself without even consulting others.

I believe that the book should serve specific need of the students appearing in B.Com. (Hons.), B.A. (Eco. Hons.), M.Com., (M.A.) (Eco.) B.B.A., B.C.A., M.B.A. and other competitive examinations.

Suggestions and comments for the improvement of the book from readers will be thankfully received and duly incorporated in the subsequent editions.

Author

CONTENTS

1

Mathematical Logics

INTRODUCTION OF LOGICS

Study of logic is very important in mathematics as it provides the theoretical basis for many areas of computer science as artificial intelligence, digital logic design etc. We shall study the algebra of statements or sentences. Also we know that a word has more than one meaning, so there is a possibility of interpreting a group of words in more than one way and thereby creating a confusion in the meaning of a statement. We use symbolic language to express mathematical statements and analysis of this symbolic language is the *Mathematical logic*. We use symbolic language to express mathematical statements and adjudge the truth or falsity of these statements through valid reasoning.

The analysis of this symbolic language may be termed as logic.

We usually denote statements by small letter p, q, r etc. It is clear that for a given statement p, exactly one of the following must hold:

(i) p is true;

(ii) p is false.

NORMAL FORMS LOGICS

By comparing truth tables, we can easily conclude whether two logical expressions P and Q are equivalent. But if the number of variables increases, the process becomes tedious. For example.

$a, a \to b, b \to c, \ldots, x \to y, y \to z \Rightarrow z$ is a theorem in propositional calculus. However, suppose that we wrote such a program and we had to write the truth table for

$$(a \land (a \to b) \land \ldots \land (y \to z)) \to z.$$

The truth table will have 2^{26} cases. At one thousand cases per second, it would take approximately 18 hours to verify the theorem.

A better method is to transform the expressions P and Q to some standard forms of expressions P' and Q' such that a simple comparison of P' and Q' shows whether P and Q are equivalent. The standard forms are called **normal forms** or **cononical forms**. There are two types of normal forms:

Dijunctive Normal Forms

In a logical expression, a product of the variables and their negations is called an **elementary product**. For example $p \wedge \sim q$, $\sim p \wedge \sim q$, $\sim p \wedge q$ are elementary products. A sum of the variables and their negations is called an **elementary sum**. For example $\sim p \vee q$, $\sim p \vee \sim q$, $p \vee \sim q$ are elementary sums. The elementary sums or products satisfy the following properties:

(i) An elementary sum is identically true if an only if it contains at least one pair of factors in which one is the negation of the other.

(ii) An elementary product is identically false if and if it contains at least one pair of factors in which one is negation of the other.

A logical expression is said to be in disjunctive normal form if it is the sum of elementary products. For example, $p \vee (q \wedge r)$ and $p \vee (\sim q \wedge r)$ are in disjunctive normal form.

Conjunctive Normal Form

A logical expression is said to be in conjunctive normal form if it consists of a product of elementary sum.

LOGICAL CONNECTIVES

Associated with every statement is another statement called its negation. In other words, the statement having meaning contradictory to the given statement p is called the negation or contradiction of p and is denoted by '~ p' *i.e.*

'~ p' ⇒ implies not p or negative of p.

It is important to note that the negation of a *true* statement is *false* and that of a *false* statement is *true*.

i.e., if p is true then ~ p is false

if p is false then ~ p is true

Example:

The negation of the statement

p: √5 is an irrational number

is ~ p: √5 is not an irrational number.

The following truth table expresses the relation between the truth values of the statement p and those of its negation.

p	~ p
T	F
F	T

In the above table p is the statement and ~ p is the contradiction or negation of p. The truth values of p are stated is the column under p and that of ~ p are stated in the column under ~ p.

Theorem :

Prove that for any statement p, ~ (~p) = p. (Law of double negation).

Proof:

We shall prove that the statements p and ~ (~p) are equivalent. Let us prepare the truth table for these two.

Truth Table

p	~ p	~ (~ p)
(1)	(2)	(3)
T	F	T
F	T	F

The table shows that if p is **T** ⇒ ~ (~p) is also T

and if p is **F** ⇒ ~ (~p) is also F.

Hence, columns (1) and (3) show that p and ~ (~p) are equivalent statements. Hence ~ (~p) = p.

Conjunction

If the compound statement is formed by connecting the two simple statements with the help of the word '**and**' then, '***and***' is called the ***conjunction*** of the statements.

Example:

If p and q are two statements such that

p: Shourya is an intelligent boy

q: Shourya will join M.B.B.S.

These two statements give rise to the compound sentence 'Shourya is an intelligent boy and he will join M.B.B.S.

In this compound sentence *and* is the conjunction for p and q.

Truth Table for p ∧ q (p and q)

If p and q denote two statements then the conjunction of p and q is denoted by p ∧ q, and is read as p and q.

The statement p ∧ q is true if and only if both p and q are true. In all other cases it is false. In other words.

if p is true and q is true then p ∧ q is true,

if p is true and q is false then p ∧ q is false,

if p is false and q is true then p ∧ q is false,

if p is false and q is false then p ∧ q is false,

This can be represented by following truth table.

Truth Table for p ∧ q

p	q	p ∧ q
T	T	T
T	F	F
F	T	F
F	F	F

Disjunction ∨ (OR)

If the compound statement is obtained by connecting the two simple statements with the help of the word *or*, then the word *or* is called the *disjunction* of the statements. If p and q are two statements, then p ∨ q denotes the **disjunction** of p and q. The symbol p ∨ q is read as p or q.

Example:

If p stands for 'I shall purchase a book' and q stands for 'I shall purchase a pencil' then p ∨ q = I shall purchase a book or a pencil.

Truth Table for p ∨ q

The statement p ∨ q will be true if either

(i) p is true, or

(ii) q is true, or

(iii) both p and q are true.

If both p and q are false, then the statement p ∨ q is false. The truth table for p ∨ q is as follows:

Truth Table for p ∨ q

p	q	p ∨ q
T	T	T
T	F	T
F	T	T
F	F	F

Implication ⇒ or Conditional Statement

A conditional statement has two simple statements connected by the word "*if, ... then*". If p and q are two statements, the word **"if, ... then"** is symbolically written as p ⇒ q. It is read as p implies q.

Example:

A father declared, "If my son stands first then I shall buy a watch for him".

The part 'If my son stands first' is called the **antecedent** and 'I shall buy a watch for him' is called the **consequent** in the above conditional statement. If we denote antecedent by p and the consequent by q then conditional statement is expressed by writing p ⇒ q. Now there are four possibilities, namely.

(i) the son stands first and his father buys a watch for him,

(ii) the son does not stand first and the father buys him a watch,

(iii) the son stands first but the father does not buy a watch for him,

(iv) the son does not stand first and the father does not buy him a watch,

It is clear that in all the four cases the father breaks his promise only in case (iii). Therefore it is the case when the conditional is false. In all other cases if is true. Thus if p is true and q is false then p ⇒ q is false otherwise it is true in all other cases. The truth table of p ⇒ q is given as follows:

Truth Table for p ⇒ q

p	q	p ⇒ q
T	T	T
F	T	T
T	F	F
F	F	T

Double Implication $\Leftrightarrow$ or Biconditional or Equivalence

A biconditional contains the connective **'if and only if'** and has two conditions, thus $p \Leftrightarrow q$ is same in the meaning as $p \Rightarrow q$ and $q \Rightarrow p$.

Example:

In the example given for conditional had the father promised "I shall buy a watch for my son if and only if he stands first", he would surely have meant that he would not buy a watch if his son does not stand first. As such in addition to case (iii), case (ii) as well would go against his promise and will make the conditional a false statement. Thus the statement $p \Leftrightarrow q$ is true if p and q have the same truth values and is false if they have the opposite truth values. The truth table for $p \Leftrightarrow q$ is as given below:

Truth Table for $p \Leftrightarrow q$

or

$(p \Rightarrow q) \wedge (q \Rightarrow p)$

p	q	$p \Leftrightarrow q$
T	T	T
F	T	F
T	F	F
F	F	T

Logically Equivalent Statements

Two statements are said to be 'logically equivalent' if both have the identical truth values, *i.e.*, in each row of the truth table both statements must have same truth values.

Logically equivalent statements are expressed by writing the symbol $\equiv$ in between or sometimes by = if there is no confusion.

Theorem (De Morgan's Law):

If p and q are two statements, then prove that

(a) $\sim (p \wedge q) = (\sim p) \vee (\sim q)$

(b) $\sim (p \vee q) = (\sim p) \wedge (\sim q)$

Proof:

(a) Let us form the truth table:

Truth Table

p (1)	q (2)	$p \wedge q$ (3)	$\sim(p \wedge q)$ (4)	$\sim p$ (5)	$\sim q$ (6)	$(\sim p) \vee (\sim q)$ (7)
T	T	T	F	F	F	F
T	F	F	T	F	T	T
F	T	F	T	T	F	T
F	F	F	T	T	T	T

In the above table columns (4) and (7) are identical and thus the statements $\sim(p \wedge q)$ and $(\sim p) \wedge (\sim q)$ are equivalent statements.

Hence $\sim(p \wedge q) = (\sim p) \vee \sim(q)$

(b) Let us construct the following truth table

Truth Table

p (1)	q (2)	$p \vee q$ (3)	$\sim(p \vee q)$ (4)	$\sim p$ (5)	$\sim q$ (6)	$(\sim p) \wedge (\sim q)$ (7)
T	T	T	F	F	F	F
T	F	T	F	F	T	F
F	T	T	F	T	F	F
F	F	F	T	T	T	T

In the above table we notice that columns (4) and (7) are identical, and therefore, the statements $\sim(p \vee q)$ and $(\sim p) \wedge (\sim q)$ are equivalent statements.

Hence $\sim(p \vee q) = (\sim p) \wedge (\sim q)$.

ARGUMENT

An argument is a statement which declares that the given set of propositions $p_1, p_2, \ldots, p_n$ yield a new proposition Q. Then argument is expressed as

$$p_1, p_2, p_3, \ldots, p_n \vdash Q$$

or $p_1 \wedge p_2 \wedge p_3 \wedge \ldots \wedge p_n \vdash Q$

The symbol $\vdash$ is known as **turnstile**. The proposition $p_1, p_2, p_3, \ldots, p_n$ are called **'premises'** (or **assumptions**) and Q is called the **'conclusion'**.

Such an argument is true, *i.e.*, valid when Q is true, *i.e.*, when all the premises $p_1, p_2, p_3, \ldots, p_n$ are true. It may so happen that p_1, p_2 being true (in the truth table) but conclusion is not true, the argument is false *i.e.*, not valid.

MATHEMATICAL SYSTEM

A mathematical system consists of:

(i) ***A set or universe U.***

(ii) ***Definitions:*** *Sentences that explain the meaning of concepts that relate to the universe are known as definitions. Any term used in describing the universe itself is said to be undefined. All definitions are given in terms of these undefined concepts of objects.*

(iii) ***Axioms:*** *Assertions about the properties of the universe and rules for creating and justifying more assertions. These rules always include the system of logic.*

(iv) ***Theorem:*** *the additional assertions mentioned above.*

Example:

In the logical system, the universe consists of propositions. The axioms are the truth tables for the logical operators and the key definitions are those of equivalence and implication.

Theorem:

A true proposition derived from axioms of mathematical system is called a theorem.

Proof:

All the theorems can be expressed in terms of a finite number of propositions, $p_1, p_2, ..., p_n$, called the **premises,** and a proposition C, called the **conclusion**. These theorems take the form

$$p_1 \wedge p_2 \wedge ... \wedge p_n \Rightarrow C$$

or more simply,

$$p_1, p_2, ..., \text{and } p_n \text{ imply } C.$$

When a theorem is stated, it is assumed that the axioms of the system are true. In addition, any previously proven theorem can be considered an extension of the axioms and can be used in demostrating that the new theorem is true. When the proof is complete, the new theorem can be used to prove subsequent theorems.

LAWS OF ALGEBRA OF PROPOSITION

Idempotent laws	$p \vee p \equiv p$	$p \wedge p \equiv p$
Associative laws	$(p \vee q) \vee r \equiv p \vee (q \vee r)$	$(p \wedge q) \wedge r \equiv p \wedge (q \wedge r)$

Commutative laws	$p \vee q \equiv q \vee p$	$p \wedge q \equiv q \wedge p$
Distributive laws	$p \vee (q \wedge r) \equiv (p \vee q) \wedge (p \wedge r)$	
		$p\wedge(q\vee r) \equiv (p\wedge q)\vee(p\wedge r)$
Identity laws	$p \vee t \equiv p$	$p \wedge t \equiv p$
	$p \vee t \equiv t$	$p \wedge f \equiv f$
Complement laws	$p \vee \sim p \equiv t$	$p \wedge \sim p \equiv f$
	$\sim \sim p \equiv p$	$\sim t \equiv f, \sim f \equiv t$
De Morgan's laws	$\sim(p \vee q) \equiv \sim p \wedge \sim q$	$\sim (p \wedge q) \equiv \sim p \vee \sim q$

PROOFS

A proof of a theorem is a finite sequence of logically valid steps that demostrate that the premises of a theorem imply the conclusion.

Proofs in Propositional Calculus (Logical System)

As deseribed earlier, a theorem is a proposition that can be proved to be true. An argument that establishes truth of a theorem is called a proof.

Valid Arguments

An argument is a sequence of statements. All statements but then final one are called premises (or assumption or hypothesis). The final statement is called conclusion.

An argument is said to be logically valid, if and only if the conjunction of the premises implies the conclusion. This means that if the premises are all true, the conclusion must also be true. However, if one or more of the premises is false, so that the conjunction of all the premises is false, then the conclusion may be either true or false.

To test the validity of an argument, the following procedure may be adopted:

(i) Identify the premises and conclusion of the argument.

(ii) Construct a truth table showing the truth values of all premises and the conclusion.

(iii) Find the rows (known as *critical rows*) in which all the premises are true.

(iv) In each critical row, determine. Whether the conclusion of the argument is also true.

(a) If in each critical row the conclusion is also true, then the argument from is valid.

(b) If there is at least one critical row in which the conclusion is false, the argument form is invalid.

Rules of Inference

The rules of inference are criteria for determining the validity of an argument. Any conclusion which is arrived by following the rules of inference is called a valid conclusion, and the argument is called valid argument. Generally, for a proof we use two fundamental rules of inference.

Rule 1: *If the statement p is assumed as true and also the statement $p \to q$ is accepted as true, then q must be true.*

Symbolically,

$$p \to q$$

$$p$$

$$\therefore q$$

In this presentation, the assertions above the horizontal line are the **premises** or **hypotheses** while the assertion below the line is the **conclusion**. The rule depicted is known as **modus ponens** or the **rule of detachment**. The validity of the argument can also seen from the truth table. For this we construct a truth table for the premises and conclusion.

Truth Table

		Premises	Conclusion	
p	q	$p \Rightarrow q$	p	q
T	T	T	T	T
T	F	F	T	F
F	T	T	F	T
F	F	T	F	F

It is clear from the truth table that there is only one case in which both premises are true (first case), and that in this case the conclusion is also true. Hence the argument is valid.

Another way of stating that the above argument is valid is that $[(p \Rightarrow q) \wedge p] \Rightarrow q$ is tautology.

Rule 2: *Whenever the two implications $p \Rightarrow q$ and $q \Rightarrow r$ are accepted as true then the implication $p \Rightarrow r$ is accepted as true.*

Symbolically:

$$p \Rightarrow q$$

$$q \Rightarrow r$$

$$\therefore \quad p \Rightarrow r$$

This argument is known as a **hypothetical syllogism.**

Truth Table

p	q	r	$p \Rightarrow q$	$q \Rightarrow r$	$p \Rightarrow r$
T	T	T	T	T	T
T	T	F	T	F	F
T	F	T	F	T	T
T	F	F	F	T	F
F	T	T	T	T	T
F	T	F	T	F	T
F	F	T	T	T	T
F	F	F	T	T	T

It is clear from the first, fifth, seventh and eighth rows of the truth table that both premises are true. Since in each case the conclusion is also true, the argument is valid.

This rule may also be described as:

$$(p \Rightarrow q) \wedge (q \Rightarrow r) \Rightarrow (p \Rightarrow r)$$

is a tautology.

Additional Valid Argument Forms

There are other valid inferences. Some of them are:

Modus Tollens

The argument of the from

$$p \Rightarrow q$$
$$\sim q$$
$$\therefore \sim p$$

This argument is called modus tollens which means "method of denying". It can easily be established by using a truth table.

Addition

The following argument form is valid.

$$p$$
$$\therefore p \vee q$$

This form is used for making generalizations. If p is true, then more generally, p or q is true for any other statement q.

Disjunctive Syllogism

The following statement form is valid

$p \vee q$

$\sim q$

$\therefore \quad p$

According to this argument, when there are two possibilities and one can rule one out, the other must be the case.

METHODS OF PROOF

Direct Proof

A direct proof is a proof in which the truth of the premises of a theorem are shown to directly imply the truth of the theorem's conclusion.

Example:

For example the direct proof of the theorem: $p \rightarrow r,\ q \rightarrow s,\ p \vee q \Rightarrow s \vee r$ *is*

Step	Proposition	Justification
(1)	$p \vee q$	Premise
(2)	$\sim p \rightarrow q$	(1) conditional rule
(3)	$q \rightarrow s$	premise
(4)	$\sim p \rightarrow s$	(2), (3), chain rule
(5)	$\sim s \rightarrow p$	(4), conditional rule
(6)	$p \rightarrow r$	premise
(7)	$\sim s \rightarrow r$	(5), (6), chain rule
(8)	$s \vee r$	(7), conditional rule

Indirect Proofs

Consider a theorem $P \Rightarrow C$, where P represents $p_1 \wedge p_2 \wedge \ldots, \wedge p_n$, the premises. The method of indirect proof is based an the equivalence $P \rightarrow C \Rightarrow \sim (P \wedge \sim C)$.

This logical law states that if $P \Rightarrow C$, then $P \wedge \sim C$ is always false, *i.e.*, $P \wedge \sim C$ is a contradiction. This means that a valid method of proof is to negate the conclusion of a theorem and add this negation to the premises. If a contradiction can be implied from this set of propositions, the proof is complete. Indirect proofs can often more convenient than direct proofs.

Conditional Conclusion

The conclusion of a theorem is often a conditional proposition. The condition of the conclusion can be included as a premise in the proof of the theorem. The object of the proof is then to prove the consequence of the conclusion. This rule is justified by the logical law

$$p \rightarrow (h \rightarrow c) \Leftrightarrow (p \wedge h) \rightarrow C$$

PROPOSITIONS OVER A UNIVERSE

Let U be a non-empty set. A proposition over U is a sentence that contains a variable that can take on any value in U and which has a definite truth value as a result of any such substitution.

Truth Set: If p(n) is a proposition over U, the truth set of p(n) is $T_{p(n)} = \{a \in U \mid p(a) \text{ is true}\}$.

Example: *The truth set of the proposition $\{a, b\} \cap A = \phi$ taken as a proposition over the power set of $\{a, b, c, d\}$ is $\{\phi, \{c\}, \{d\}, \{c, d\}\}$.*

Tautology and Contradiction: *A proposition over U is a tautology if its truth set is U. It is a contradiction if its truth set is empty.*

Example: *$(p + 1)(p - 1) = p^2 - 1$ is a tautology over the rationals, $x^2 - 3 = 0$ is a contradiction over the rationals.*

Equivalence: Two propositions are equivalent if $p \Rightarrow q$ is a tautology. In terms of truth sets, this means that p and q are equivalent if $T_p = T_q$.

Example: *$n + 3 = 7$ and $n = 4$ are equivalent propositions over the integers.*

Implification: If p and q are propositions over U, p implies q if $p \rightarrow q$ is a tautology. Since the truth set of $p \rightarrow q = T_p' \cap T_q$, then $p \Rightarrow q$. When $T_p \subset T_q$.

Example: Over the natural numbers: $n \leq 3 \Rightarrow n \leq 6$ since $\{0, 1, 2, 3,\} \subset \{0, 1, 2, 3, 4, 5, 6,\}$.

TAUTOLOGY AND FALLACY

A statement is said to be a prime statement if it does not contain any connectives.

Tautology: A statement is said to be a tautology if its truth value is T irrespective of the truth or falsity of its prime statements. A tautology is generally denoted by t.

Fallacy or Contradiction: A statement is said to be a contradiction or a fallacy if its truth value is F irrespective of the truth or falsity of its prime statements. A fallacy is generally denoted by f.

Remark: It is obvious from above definitions that if a statement is a fallacy then its negation is a tautology and vice - versa.

Logically True Statement: A statement is logically true if it is derivable from a tautology. For example, the statement is $p \vee \sim p$ is a tautology and as such it is logically true.

SIMPLE STATEMENT

A statement p is said to be *simple* if it has one subject and one predicate.

COMPOUND STATEMENT

A statement is said to be *compound statement* if it is formed of two or more than two simple statements. The simple statements are called the components of the compound statement. The compound statement is formed by connecting the simple statements with the help of the words **'and' 'or'**, **'if'** ... **'then'**.

Example:

If p and q are two statements such that

p: Gunjan is an intelligent girl

q: Gunjan will join M.C.A.

Now, p and q are two simple sentences and they can be combined in a number of ways to form a *Compound* sentences such as:

(i) Gunjan is an intelligent *girl and* she will join M.C.A.

(ii) Gunjan is intelligent *or she* will join M.C.A.

(iii) If Gunjan is intelligent *then* she will join M.C.A.

(iv) Gunjan is an intelligent *girl if and only if* she joins M.C.A.

TRUTH VALUE OF A STATEMENT

We know that every statement is a sentence, which is either true or false. The truth or falsity of a statement is called its *truth value*. We assign to the statement p, the letter **T** when p is true and the letter **F**, when it is false. Both **T** and **F** are called the *truth values* of the statement p.

EQUIVALENT STATEMENTS

If the truth values of two statements are identical, then they are equivalent, *i.e.*, if p and q are two statements, then p and q are equivalent statements if p is **T** and q is also **T** or if p is **F** and q is also **F**. The equivalent statements are denoted by $p = q$ or $p \Leftrightarrow q$.

Example:

Let the statements p and q be as follows:

p: n is an even integer

q: n + 1 is an odd integer.

Now, if n is an even number, then we know that n + 1 is an odd number, therefore p $\Leftrightarrow$ q. Hence p and q are equivalent statements.

SOLVED EXAMPLES

Example 1:

In a certain country it is found that weather follows the following rules:

If it is fine today, then it is windy tomorrow

If it is calm today, then it is hot tomorrow

If it is fine tomorrow, then it is cold tomorrow

Each day is either hot or cold, wet or fine and calm or windy. Forecast tomorrow's weather if today is fine, calm and cold.

Solution:

The argument is as follows:

(i) fine today $\Rightarrow$ windy tomorrow

today is fine, $\therefore$ tomorrow is windy

(ii) calm today $\Rightarrow$ hot tomorrow

today is calm, $\therefore$ tomorrow is hot

The truth table for p $\Rightarrow$ q and ~q $\Rightarrow$ ~p are same, so that proposition p $\Rightarrow$ q and ~q $\Rightarrow$ ~p are equivalent. In other words p $\Rightarrow$ q $\equiv$ ~q $\Rightarrow$ ~p.

$\therefore$ fine tomorrow

$\Rightarrow$ cold tomorrow

$\equiv$ ~(cold tomorrow) $\Rightarrow$ ~ (fine tomorrow)

$\equiv$ hot tomorrow $\Rightarrow$ wet tomorrow

(iii) hot tomorrow $\Rightarrow$ wet tomorrow

tomorrow is hot, $\therefore$ tomorrow is wet.

The forecast is hot, wet and windy.

Example 2:

Prove that proposition ~[p $\wedge$ (~p)] is a tautology.

Solution:

Let us prepare the truth table for the given proposition.

Truth Table

p	~p	p ∧ (~p)	~[p ∧ (~p)
(1)	(2)	(3)	(4)
T	F	F	T
F	T	F	T

Since the column (4) contains T everywhere, therefore this proposition is a tautology.

Example 3:

Represent the argument.

If it rains today, then we will not play cricket today.

If we don't play cricket today, then we will play cricket tomorrow. ....

Therefore, if it rains today, then we will play cricket tomorrow,

symbolically and determine whether the argument is valid.

Solution:

Let

p: It is raining today

q: We will not play cricket today

r: We will play cricket tomorrow.

The argument is of the form

$$p \Rightarrow q$$

$$q \Rightarrow r$$

$$\therefore\ p \Rightarrow r$$

Hence the argument is a hypothetical syllogism and thus the argument is valid.

Example 4:

Represent the argument.

If this number is divisible by 4, then it is divisible by 2.

This number is not divisible by 2. ...

This number is not divisible by 4.

symbolically and determine whether the argument is valid.

Solution:

Let

p: The number is divisible by 4.

q: It is divisible by 2.

The argument may be written as

$$p \Rightarrow q$$
$$\sim q$$
$$\therefore \quad \sim p$$

Thus by modus tollens the argument is valid

Example 5:

Represent the argument.

Either Gunjan is not guilty or Rashmi is telling the truth. ...

Rashimi is not telling the truths

Therefore Gunjan is not guilty.

symbolically and determine whether the argument is valid.

Solution:

Let

p: Gunjan is not guilty

q: Rashmi is telling the truth

The argument can be written as

$$p \vee q$$
$$\sim q$$
$$\therefore \quad p$$

Thus by disjunctive syllogism, the argument is valid.

Example 6:

Prove that s is a valid conclusion from the premises $p \Rightarrow q$, $p \Rightarrow r$, $\sim (q \wedge r)$ *ad* $s \vee p$.

Solution:

To prove the theorem we have the following table:

1. $p \Rightarrow q$ Premise (given)
2. $p \Rightarrow r$ Premise (given)

3. $(p \Rightarrow q) \wedge (p \Rightarrow r)$ From 1 and 2
4. $\sim(q \wedge r)$ Premise given)
5. $\sim q \vee \sim r$ De Morgan's law from 4
6. $\sim p \vee \sim p$ Using 3 and 5
7. $\sim p$ Idempotent law from 6
8. $s \vee p$ Premise (given)
9. s Disjunctive syllogism from 7 and 8

Thus s is valid from the given premises.

Example 7:

Prove the validity of the following argument "If I get the admission and work hard then I will get first division. If I get first division, then I will be happy. I will not be happy.

Therefore, either I will not get the admission or I will not work hard".

Solution:

Let

p: I get the admission

q: I work hard

q: I get first division

s: I will be happy.

Then the above argument can be written in symbolic for as

$$(p \wedge q) \Rightarrow r$$
$$r \Rightarrow s$$
$$\sim s$$

Therefore,

1. $(p \wedge q) \Rightarrow r$ Premise (given)
2. $r \Rightarrow s$ Premise (given)
3. $(p \wedge q) \Rightarrow s$ Hypothetical syllogism by 1 and 2
4. $\sim s$ Premise (given)
5. $\sim(p \wedge q)$ Modus tollens by 3 and 4
6. $\sim p \vee \sim q$ Conclusion.

Hence the argument is valid.

Example 8:

Prove that product of two odd integers is an odd integer.

Solution:

Let p and q are two odd integers. Then there exist two integers m and n, so that $p = 2m + 1$ and $q = 2n + 1$. Then

$$pq = (2m + 1)(2n + 1) = 4mn + 2m + 2n + 1$$
$$= 2(2mn + m + n) + 1 \text{ which is odd.}$$

Example 9:

Prove that $\sqrt{3}$ is irrational by giving a proof by contradiction.

Solution:

Let $\sqrt{3}$ is a rational. Under assumption that $\sqrt{3}$ is rational, there exist integers p and q such that $\sqrt{3} = p/q$. Where p and q have no common factors. Squaring both sides, we get

$$3 = p^2/q^2 \Rightarrow 3q^2 = p^2$$

Hence p^2 is a multiple of 3, and therefore odd. This implies p is even. Hence $p = 3k$ for some integer k. Then $3q^2 = (3k)^2 \Rightarrow q^2 = 3k^2$. Thus q^2 is odd, q is odd. But now p and q have a common factor of 3, which is a contradiction to the statement that p and q have no common factors.

Hence our initial assumption that $\sqrt{3}$ is rational is false. Thus $\sqrt{3}$ is irrational.

Example 10:

Prove that for every positive integer n, $n^3 + n$ is even.

Solution:

When n is even: Then $n = 2k$ for some positive integer k.

Now $\quad n^3 + n = (2k)^3 + 2k = 8k^3 + 2k$

$$= 2(4k^3 + k) \text{ which is even.}$$

When n is odd: Then $n = 2k + 1$ for some positive integer k.

Now $\quad n^3 + n = (8k^3 + 12k^2 + 6k + 1) + (2k + 1)$

$$= 8k^3 + 12k^2 + 8k + 2$$
$$= 2(4k^3 + 6k^2 + 4k + 1)$$

Which is even

Hence the sum $n^3 + n$ is even.

Example 11:

Simplify the following statements:

(i) $\sim \sim p$,

(ii) $\sim(p \vee \sim q)$,

(iii) $\sim(\sim p \wedge q)$,

(iv) $\sim(\sim p \vee \sim q)$,

(v) $(p \vee q) \wedge \sim p$

Solution:

(i) $\sim \sim p = p$ ($\therefore$ negative of a negative is positive)

(ii) $\sim(p \vee \sim q) = \sim p \wedge \sim \sim q = \sim p \wedge q$ ($\because \sim\sim q = q$)

(iii) $\sim(\sim p \wedge q) = \sim \sim p \vee \sim q = p \vee \sim q$ ($\because \sim \sim p = p$)

(iv) $\sim(\sim p \vee \sim q) = \sim \sim p \wedge \sim \sim q = p \wedge q$

(v) $(p \vee q) \wedge \sim p = \sim p \wedge (p \vee q)$ (Commutative law)

$= (\sim p \wedge p) \vee (\sim p \vee q)$ (Distributive law)

$= f \vee (\sim p \wedge q) = \sim p \wedge q.$

Example 12:

Simplify the following statements:

(i) $p \vee (p \wedge q)$ *(ii)* $\sim(p \vee q) \wedge (\sim p \wedge q)$

Solution:

(i) $p \vee (p \vee q) = (p \wedge t) \vee (p \wedge q)$ (Identity law) ($\because p = p \vee t$)

$= p \wedge (t \vee q)$ (Distributive law)

$= p \wedge t = p$ ($\because t \vee q = t$)

(ii) $\sim (p \vee q) \vee (\sim p \wedge q)$

$= (\sim p \wedge \sim q) \vee (\sim p \wedge q)$ (De Morgan's law)

$= \sim p \wedge (\sim q \vee q)$ (Distributive law)

$= \sim p \wedge t = \sim p$ (Complement law)

Example 13:

Show that the statement $(p \wedge q) \Rightarrow p$ *is a tautology.*

Solution:

Let us prepare the truth table for the statement $(p \wedge q) \Rightarrow p$.

Truth Table

p	q	p ∧ q	(p ∧ q) ⇒ p
(1)	(2)	(3)	(4)
T	T	T	T
T	F	F	T
F	T	F	T
F	F	F	T

In the above table we notice that column (4) under (p ∧ q) ⇒ p has all its entries as T.

Hence (p ∧ q) ⇒ is a tautology.

Example 14:

Prove that the statement [(p ⇒ q) ∧ (q ⇒ r)] ⇒ (p ⇒ r) is a tautology

Solution:

Let us construct the truth table for the given proposition.

Truth Table

p	q	r	(p⇒q)	(q⇒r)	(p⇒r)	[p⇒ r∧q ⇒r]	[p⇒q∧q⇒r] ⇒ p ⇒ r
(1)	(2)	(3)	(4)	(5)	(6)	(7)	(8)
T	T	T	T	T	T	T	T
T	T	F	T	F	F	F	T
T	F	T	F	T	T	F	T
T	F	F	F	T	F	F	T
F	T	T	T	T	T	T	T
F	T	F	T	F	T	F	T
F	F	T	T	T	T	T	T
F	F	F	T	T	T	T	T

Since the column (8) contains all its entries as T, therefore, the given proposition is a tautology.

Example 15:

Show by means of a truth table that ~(p ⇒ q) = p ∧ ~q

Solution:

Let us construct the truth table for the given proposition

$$\sim (p \Rightarrow q) = p \wedge \sim q$$

Truth Table

p (1)	q (2)	$p \Rightarrow q$ (3)	$\sim(p \Rightarrow q)$ (4)	$\sim q$ (5)	$p \wedge \sim q$ (6)
T	T	T	F	F	F
T	F	F	T	T	T
F	T	T	F	F	F
F	F	T	F	T	F

The given statement is valid as the truth values under column (4) and column (6) are alike. Hence $\sim (p \Rightarrow q) = p \wedge \sim q$.

Example 16:

Prove by means of a truth table

$$p \Rightarrow (q \wedge r) = (p \Rightarrow q) \wedge (p \Rightarrow r)$$

Solution:

Let us construct the truth table for the given proposition.

Truth Table

p (1)	q (2)	r (3)	$q \wedge r$ (4)	$p \Rightarrow (q \wedge r)$ (5)	$p \Rightarrow q$ (6)	$p \Rightarrow r$ (7)	$(p \Rightarrow q) \wedge (p \Rightarrow r)$ (8)
T	T	T	T	T	T	T	T
T	T	F	F	F	T	F	F
T	F	T	F	F	F	T	F
T	F	F	F	F	F	F	F
F	T	T	T	T	T	T	T
F	T	F	F	T	T	T	T
F	F	T	F	T	T	T	T
F	F	F	F	T	T	T	T

Hence we notice that the truth values of columns (5) and (8) are alike, therefore the two statements are equivalent.

Hence, $p \Rightarrow (q \wedge r) = (p \Rightarrow q) \wedge (p \Rightarrow r)$

Example 17:

By means of truth table, prove that

$$p \Rightarrow q = (p \Rightarrow q) \wedge (q \Rightarrow p)$$

Solution:

Let us construct the truth table for the given proposition.

Truth Table

p	q	$p \Leftrightarrow q$	$p \Rightarrow q$	$q \Rightarrow p$	$(p \Rightarrow q) \wedge (q \Rightarrow p)$
(1)	(2)	(3)	(4)	(5)	(6)
T	T	T	T	T	F
T	F	F	F	T	F
F	T	F	T	F	F
F	F	T	T	T	T

The truth values of columns 3 and 6 are alike, so the given proposition is true.

EXERCISES

1. Prove that the proposition $[\sim p \wedge (\sim p)]$ is a tautology.
2. Show that the statement $p \wedge \sim p$ is a fallacy.
3. Prove that the sentence "It is raining or it is not raining" is a tautology.
4. By means of truth table, prove that

   $$p \vee q \equiv (p \vee q) \wedge \sim (p \wedge q)$$

5. By means of a truth table, prove that

   $$p \wedge (q \vee r) \equiv (p \wedge q) \vee (p \wedge r)$$

6. A tautology is a statement which is always true. Using truth tables, show that $(p \wedge q) \Rightarrow p$ and $p \Rightarrow (p \vee q)$ are both tautologies where p, q are any two statements.
7. Show that the statement

   $$p \vee (q \wedge r) \Leftrightarrow (p \vee q) \wedge (p \vee r),$$

   where p, q, r are statements, is a tautology.
8. Prove that:

   (a) $p \Rightarrow q = (\sim p) \Rightarrow (\sim q)$

(b) $\sim (p \Rightarrow q) = p \wedge (\sim q)$

(c) $p \Rightarrow q = (\sim p) \vee q$

(d) $[(p \rightarrow q) \wedge \sim q] \rightarrow \sim p$.

9. Show that $t \Rightarrow s$ is a valid conclusion from the given premises

$(p \wedge q) \vee (r \Rightarrow s)$,

$t \Rightarrow r, \sim (p \wedge q)$

10. Show that s is a valid conclusion from the given premises

$p \Rightarrow \sim q, \quad q \vee r,$

$\sim s \Rightarrow p, \quad \sim r,$

2

Set Theory

INTRODUCTION

As a matter of fact, pure mathematics can be described as a study of set equipped with assigned structures, known as mathematical systems.

Definition: *A set as a well defined collection of objects* by the term *'well* **defined'**, we mean that we must given a rule with the help of which we should able to tell whatever a given object belongs or does not belongs to that given collection.

The object in a set are called its members or elements. We usually denote sets by capital letters and their elements by small letters.

If an object x is a member of a set A, we write $x \in A$, which means that 'x belongs to A' or that 'x is an element of A'. On the other hand, if x does not belong to A, we write $x \notin A$.

Example of Sets

(i) The collection of vowels in English alphabet is a set, containing five elements, namely a, e, i, o, u.

(ii) The collection of four fourth roots of unity, constitutes a set with 1, –1, i, –i as its elements,

(iii) The collection of first four counting numbers is a set containing 1, 2, 3, 4.

Note: The terms good, bad, rich, poor, honest etc. are vague and these are not well defined. Thus.

HOW TO DESCRIBE OR SPECIFY A SET

There are two methods for describing a set:

(1) **Tabulation of Roaster Method:** *Under this method we just make a list of the objects of the set and put it within braces { }.* or

In tabulation each element is seperated by commas and inclosed in brackets.

Examples:

(i) If B is the set of four fourth roots of unity, then B = {1, –1, i, –i}.

(ii) If A is the set of vowels in English alphabet, then A = {a, e, i, o, u}

(iii) **Description or Set Builder Method:** *This method consists in the listing of the property or properties satisfied by the elements of the set.* We write, {x: x *satisfied Properties P}, i.e.,* the set of all those elements such that each element x satisfies the properties P.

Examples:

(i) If B = {3, 5, 7, 9, 11}, then we may write.

B = {x: x = 2n + 1, where n is a counting number, n < 6}.

(ii) If A is the set of first four counting numbers, then A = {x: x is a counting number, x < 5}.

SETS OF NUMBERS

(a) **Natural Numbers:** *Counting numbers are called natural numbers.* Thus, the set N = {1, 2, 3, 4, ...} is the set of all natural numbers. Note that $0 \notin N$

(b) **Integers:** *The set Z consisting of all natural numbers, 0 and negatives of natural numbers, is the set of all integers.*

Thus, Z = {0, –3, –2, –1, 0, 1, 2, ...}.

Remarks:

(a) The integers divisible by 2 are *even integers,* while the integers not divisible by 2 are *odd integers.*

(b) The set of *+ve integers is,* Z^+ = {1, 2, 3, 4, ...}

The set of *–ve integers is,* Z^- = {–1, –2, –3, ...}

Note: 0 is neither +ve nor –ve.

(c) **Whole Numbers:** *The set obtained by adjoining 0 to the set of all natural numbers, is the set W of* ***whole numbers****.*

Thus, W = {0, 1, 2, 3, 5}.

(d) **Real Number:** *The totality of retionals and irrationals forms the set R of all* **real numbers.**

(e) **Rational Numbers:** *The set Q of all numbers of the form (p/q) where p and q are integers and $q \neq 0$, is the set of* ***rationals.***

Thus, Q = *{p/q: p and q are integers q ≠ 0}*.

(f) **Irrational Numbers:** *The set of all numbers which can be expressed in decimal form in non-terminating and non-repeating form only, is the set of irrationals. e.g.,* √2, √3, √5, π, e etc. are irrationals.

(g) **Intervals:** If a and b are real numbers such that a , b, then

(a) the set { x ∈ R: a ≤ x ≤ b} is called a *closed interval* denoted by [a, b];

(b) the set {x ∈ R: a < x < b} is called an *open interval* denoted by]a, b[;

(c) the set {x ∈ R: a < x ≤ b} is called a *left half open interval*, denoted by]a, b]; and

(d) the set {x ∈ R: a ≤ x < b} is called a *right half open interval*, denoted by [a, b[.

(h) **Complex Number:** The set

Z = {a + ib: a and b are real and i = √–1 is the set of all *Complex Numbers.*

EQUIVALENT SETS

The number of distinct element contained by a finite set A is called the **cardinal number** *of A, to be denoted by n (A). Two finite sets are said to be* **equivalent,** *if they have the same number of distinct elements.*

Remark:

Equivalent sets are not always equal, but equal sets are always equivalent e.g. {1, 2, 3} and {a, b, c} are equivalent, but not equal.

SUBSETS

If A and B be two sets given in such a way that every element of A is in B, then we say that A is a **subset** *of B and write,* ***A ⊆ B****. Also, if* ***A ⊆ B****, then we say that B is a* **superset** *of a and write,* ***B ⊇*** *A. If* ***A ⊆ B*** *but* ***A ≠ B****, then A is called a* **proper subset** *of B, written as,* ***A ⊂ B.***

Two sets A and B are said to be comparable if A is the subset of B or B is the subset of A, that is if A ⊆ B of B ⊇ A.

Notes:

(i) Every set is a subset of itself.

(ii) If ∃ even a single element in A which is not in B, then A is not a subset of B and we write, A ⊄ B.

(iii) If A and B are two sets given in such a way that no element is common to both A and B, then A and b are said to be *disjoint* sets.

(iv) ϕ has no proper subset.

Examples:

(i) $N \subset W \subset Z \subset Q \subset R \subset C$ but $W \not\subseteq Z^+$, $\{0\} \not\subseteq Z^+$,

(ii) $\{x \in z$: x is a multiple of $6\} \subset \{x \in Z$: x is a multiple of $3\}$

(iii) $\{1, 2\} \not\subset \{2, 4, 6, 8\}$

(iv) $\{1\} \subset \{1, 2, 3\}$.

Axiom of Specification: If A is a set and P(x) is a statement then there exist a subset B of A whose elements are exactly those elements x of A for which the statement P(x) is true.

We say that the subset B is specified by the statement P(x) notice that this subset B is unique statement P(x), then by the axim of identity B = B'.

We write $B = \{x: P(x) . x \in A\}$,

or simply $B = \{x: P(x)\}$.

Example 1:

(i) Let A = {x: x is even}, *i.e.*, A = {2, 4, 6, . . .) and let B = {x: x is a positive poweer of 2}, *i.e.*, B = {2, 4, 8, 16}. Then $A \subset B$, *i.e.*, A is contained in B.

(ii) If P be the set of all parallelograms and S is the set of all squares in a plane, *i.e.*,

P = {all parallelograms in the plane}

and S = {all squares in the plane}

Then S is a subset of P, *i.e.*,

$S \subset P$.

Remark 1: *If we have to prove $A \subset B$, then we should prove that $x \in A \Rightarrow x \in B$.*

Symbolically, $A \subset B$ iff $(x \in A \Rightarrow x \in B)$.

Remark 2: *If we have to prove that $A \not\subset B$ then we should show that there exists atleast one element x such that $x \in A$ but $x \notin B$.*

Symbolically, $A \not\subset B$ iff $\{\exists\, x \in A$ s.t. $x \notin B)$.

The symbol $\exists$ stands for 'there exists' and s.t. for 'such that'.

Example 2:

Write the subset of the following set:

$$A = \{1, 2, 3\}.$$

Solution:

The required subsets are:

$$\phi, \{1\}, \{2\}, \{3\}, \{1, 2\}, \{1, 3\}, \{2, 3\}, \{1, 2, 3\}.$$

Equality of Sets: Set A is said to be equal to set B if both of them have the same members, *i.e.*, every element which belongs to A also belongs to B ($A \subset B$) and every element which belongs to B also belongs to A ($B \subset A$). We denote the equality of sets A and B by

$$\Rightarrow A = B.$$

Proper Subsets: Let A and B be two sets. If $B \subset A$ and $B \neq A$, then B is said to be a proper subset of A.

For example, if A {1, 2, 3} and B = { 2, 3} then B is proper subset of A because all the elements of B are in A but one element 1 of A is not in B.

Theorem 1:

Every set is a subset of itself.

Proof:

Let A be any set. Then each element of A is clearly in A itself. Hence $A \subset A$.

Theorem 2:

If $A \subset B$ and $B \subset C$ then $A \subset C$.

Proof:

We must show that every element in A is also an element is C. Let x be an element of A, *i.e.*,

$$x \in A.$$

But it is given that $A \subset B$.

$$\therefore \quad x \in A \Rightarrow x \in B \qquad ...(1)$$

Also $\quad B \subset C$

$$\therefore \quad x \in B \Rightarrow x \in C \qquad ...(2)$$

Hence from (1) and (2)

$$x \in A \Rightarrow x \in C$$

$$\therefore \quad A \subset C.$$

Theorem 3:

Null set ϕ is a subset of every set.

Proof:

Let A be any given set. We shall prove that $\phi \subset A$. Let us suppose, on contrary, that $\phi \not\subset A$.

$\phi \not\subset A \Rightarrow$ there is atleast one element x such that $x \in \phi$

and $x \notin A$ ...(1)

But since ϕ is a null set,

therefore $x \notin \phi$ ...(2)

Therefore, (1) and (2) $\Rightarrow x \in \phi$ and $x \notin \phi$, which is absurd and therefore, our assumption is wrong. Consequently $\phi \subset A$.

Since A is an arbitrary set ϕ is a subset of every set.

Number of Subsets of a Finite Set: Consider a set {a}. It has two possible subsets {a} and ϕ. Again for the set {a, b}, the possible subsets are ϕ, {a}, {b}, {a, b} which are 4 subsets in number. A set {a, b, c} has as many as the following subsets.

ϕ, {a}, {b}, {c}, {a, b},

{b, c}, {a, c}, {a, b, c}.

which are eight in numbers.

Thus we can generalize the result as follows:

A set with 1 element has 2^1 subsets.

A set with 2 element has 2^2 subsets.

A set with 3 element has 2^3 subsets.

Theorem 4:

The total number of subsets of a finite set containing n elements is 2^n.

Let A be a finite set containing n elements. Let $O \leq r \leq n$. Consider those subsets of A that have r elements each. We know that the number of ways in which r elements can be chosen out of n elements is ${}^nC_r$. Therefore, the number of subsets of A having r elements each is ${}^nC_r$.

Hence, the total number of subsets of A.

$$= {}^nC_0 + {}^nC_1 + {}^nC_2 + \ldots + {}^nC_r + \ldots + {}^nC_n$$

$$= (1 + 1)^n = 2^n.$$

SOME RESULTS ON SUBSETS

Theorem 1:

The total number of subsets of a given set containing n elements is 2^n.

Proof:

Let A be an arbitrary set containing n elements. Then, one of its subsets is the empty set. Apart from this,

the number of singleton subsets of A = n = ${}^nc_1$,

the number of subsets of A, each containing 2 elements = ${}^nc_2$,

the number of subsets of A, each containing 3 elements = ${}^nc_3$,

....

the number of subsets of A, each containing (n – 1) elements = ${}^nc_{n-1}$,

the number of subsets of A, each containing n elements = ${}^nc_n$,

total number of subsets of A

$$= 1 + {}^nc_1 + {}^nc_2{}^nc_3 + \dots + {}^nc_{n-1} + {}^nc_n,$$

$$= (1 + 1)^2 = 2^n \text{ [Using Binomial Theorem]}$$

Theorem 2:

The empty set is a subset of every set.

Proof:

Let A be an arbitrary set. Then, in order to show that $\phi \subset A$, we must show that there is no element of ϕ which is not contained in A and since ϕ contains no element at all, no such element can therefore be found. Hence $\phi \subset A$.

CARTESIAN PRODUCT OR DIRECT PRODUCT OF SETS

If A and B are two non-empty sets, then the set of all ordered pairs (a, b) such that $a \in A$ and $b \in B$ is called the Cartesian product of A and B, to be denoted by $A \times B$.

Thus $A \times B = \{(a, b): a \in A \text{ and } b \in B\}$

In general, $A \times B \neq B \times A$.

Example:

If A = {0, 1} and B = {1, 2, 3}, then we have

$A \times B = \{(0, 1)\ (0, 2)\ (0, 3)\ (1, 1)\ (1, 2)\ (1, 3)\}$

$B \times A = \{(1, 0)\ (1, 1)\ (2, 0)\ (2, 1)\ (3, 0)\ (3, 1)\}$

Clearly, $A \times B \neq B \times A$.

Notes:

It may be noted that:

(i) If either A or B is an infinite set and the another one is a non-empty set, the $A \times B$ is an infinite set.

(ii) If $A = \phi$ or B ϕ, then $A \times B = \phi$.

(iii) If A has m elements and B has n elements, then $A \times B$ has mn elements.

Like ordered pairs we may define ordered triples as a set of three elements listed in a specific order and thus we define, $A \times B \times C = \{(a, b, c) : a \in A, b \in B \text{ and } c \in C\}$.

In a similar fashion, we define an n-tuple as a set of n elements listed in a specific order as $(a_1, a_2,..., a_n)$. Thus the product of n sets may be defined as,

$$A_1 \times A_2 \times ...A_n = \{(a_1, a_2,..., a_n): a_1 \in A_i \text{ for } i = 1, 2,..., n\}.$$

SOME RESULTS ON CARTESIAN PRODUCT OF SETS

Theorem 1:

For any three sets A, B and C we have

(i) $A \times (B \cup C) = (A \times B) \cup (A \times C)$.

(ii) $A \times (B \cap C) = (A \times B) \cap (A \times C)$.

(iii) $A \subseteq B \Rightarrow A \times C \subseteq B \times C$.

(iv) $A \times (B - C) = (A \times B) (A \times C)$.

Proof:

(i), (ii), (iii) Do yourself. (iv) Let (a, b) be an arbitrary element of $A \times (B - C)$. Then

$$(a, b) \in A \times (B - C)$$
$$\Rightarrow a \in A \text{ and } b \in (B - C)$$
$$\Rightarrow a \in A \text{ and } (b \in B \text{ and } b \notin C)$$
$$\Rightarrow (a \in A \text{ and } b \in B) \text{ and } (a \in A \text{ and } b \notin C)$$
$$\Rightarrow (a, b) \in (A \times B) \text{ and } (a, b) \notin (A \times C)$$
$$\Rightarrow (a, b) \in (A \times B) - (A \times C)$$
$$\therefore A \times (B - C) \subseteq (A \times B) - (A \times C)$$

Similarly, $(A \times B) - (A \times C) \subseteq A \times (B - C)$

Hence $A \times (B - C) = (A \times B) - (A \times C)$.

Theorem 2:

For any sets A, B, C and D we have

(i) $(A \times B) \cap (C \times D) - (A \cap C) \times (B \cap D)$.

(ii) $A \subseteq B$ and $C \subseteq D \Rightarrow (A \times C) = (B \times D)$

Proof:

Do yourself.

Theorem 3:

If A and B are two non-empty sets having n elements in common, then A × B and B × A have n^2 elements in common.

Proof:

Let $C = A \cap B$. Then, we claim that

$C \times C = \{A \times B) \cap (B \times A)$, Since

$(a, b) \in C \times C$

$\Leftrightarrow$ $a \in C$ and $b \in C$

$\Leftrightarrow$ $a \in A \cap B \in A \cap B$ $[\because C = A \cap B]$

$\Leftrightarrow$ $(a \in A \;\&\; b \in B)$ and $(a \in B \;\&\; b \in A)$

$\Leftrightarrow$ $(a, b) \in (A \times B)$ and $(a, b) \in (B \times A)$

$\Leftrightarrow$ $(a, b) \in (A \times B) \cap (B \times A)$.

Thus $C \times C\ (A \times B) \cap (B \times A)$.

And, since $C \times C$ has n^2 elements, so $(A \times B) \cap (B \times A)$ has n^2 elements *i.e.,* A × B and B × A have n^2 elements in common.

RELATIONS

Let A and B be two non-empty sets. Then, a relation from A to B is a subset of A × B. Thus.

R is a relation from A to B $\Leftrightarrow R \subseteq (A \times B)$.

It A = B we say that R is the relation on A we write a Rb iff $(a, b) \in R$ and say that a is related to b or that b is relative of a.

We also write a (~R)b when a is not R-relative to b.

It a consist of m elements and B consist of n elements then A × B consist of m × n elements and so (P(A × B) will have 2^{mn} elements hence the total number of different relation from A to B is 2^{mn}.

Notes:

(i) Let R $\subseteq$ A $\times$ B be a relation from A to b. Now, if (a, b) $\in$ R, then we say that 'a is R – related to b', and we write, a R b.

On the other hand, if (a, b) $\notin$ R, then 'a is not R-related to b', and we write, a K b.

(ii) The sets A $\times$ B and ϕ being subsets of A $\times$ B, it follows that each one is a relation from A to B. More-over, A $\times$ B is known as a universal relation, while ϕ is called an empty relation.

Domain and Range of a Relation: If R is a relation from a set A to another set B. Then, we domain of R is the set of all first element of R and the range of R is the set of all second coordinate of the member of R.

Domain (R) = {a: (a, b) $\in$ R} and **range (R)** = {b: (a, b) $\in$ R}.

Example:

Let A = {1, 2, 3,} and B = {2, 4, 5}, then

A $\times$ B = {(1, 2), (1, 4), (1, 5), (2, 2), (2, 4), (2, 5), (3, 2), (3, 4), (3, 5)}

Let R = {1, 2), (1, 5), (3, 4)}.

Then, R being a subset of A $\times$ B, it is a relation from A to B.

Here (1 R 2), (1 R 5), (3 R 4).

Apart from these, no element of A is related to an element of B.

Clearly, 2 R 1, 3 R 1, 4 R 2 etc.

Here, **Dom (R)** = {1, 3} and **range (R)** = {2, 4, 5}.

The Indentity Relation: Let A be any set, then the identity relation IA or (diagonal relation) on A is defined by setting (a, b) IA iff a = b the domain and range of IA are both A.

Example:

Let A = {1, 2, 3, 4} then the identity relation IA = {(1, 1), (2, 2), (3, 3), (4, 4)}

INVERSE RELATION

Let R $\subseteq$ A $\times$ B be a relation from A to B. Then, the inverse of R, denoted by R^{-1} is a relation from B to A, defined as

R^{-1} = {b, a): $\in$ R}.

Clearly, the domain and range of R^{-1} are respectively the range and domain of R.

Example:

Let A {0, 1, 2} and B {1, 3, 4}.

Now, If R = {(0, 1), (1,1), (1, 3), (2, 3)},

Then R^{-1} = {(1, 0), (1, 1) (3, 1), (3, 2)}.

Clearly, dom (R^{-1}) = {1, 3} = range (R)

and range (R^{-1}) = {0, 1, 2} = dom (R).

COMPOSITE RELATION

Let R and S be the relations from the sets A to B and B to C respectively. Then, the composite of R and S is a relation from A to C, denoted by SoR and defined by

SoR = {(a, c): $\exists$ b $\in$ B such that (a, b) $\in$ R and (b, c) $\in$ S}.

Thus, (a, b) $\in$ R, (b, c) $\in$ S $\Rightarrow$ (a, c) $\in$ SoR.

Example:

If two relations R and S be given as

R = {(1, 3), (2, 1), (3, 4), (4, 2)}

and S = {1, 2), (2, 3), (3, 4), (4, 1)}

find SoR

Solution:

Clearly, dom (SoR) = dom R.

Now (1, 3) $\in$ R, (3, 4) $\in$ S $\Rightarrow$ (1, 4) $\in$ SoR;

(2, 1) $\in$ R, (1, 2) $\in$ S $\Rightarrow$ (2, 2) $\in$ SoR;

(3, 4) $\in$ R, (4, 1) $\in$ S $\Rightarrow$ (3, 1) $\in$ SoR;

(4, 2) $\in$ R, (2, 3) $\in$ S $\Rightarrow$ (4, 3) $\in$ SoR;

$\therefore$ SoR = {(1, 4), (2, 2), (3, 1), (4, 3)}.

Theorem:

Let A, B and C be non empty sets and let $R \subseteq A \times B$ and $S \subseteq B \times C$, then $(SoR)^{-1} = (R^{-1}oS^{-1})$.

Proof:

By definition we have

$R \subseteq A \times B$, $S \subseteq B \times C \Rightarrow SoR \subseteq A \times C \Rightarrow (SoR)^{-1} \subseteq C \times A$.

Let (c, a) be an arbitrary element of $(SoR)^{-1}$ Then

$(c, a) \in (SoR)^{-1}$

$\Rightarrow$ $(a, c) \in (SoR)$

$\Rightarrow$ $\exists\, b \in B$ such that $(a, b) \in R$ and $(b, c) \in S$

$\Rightarrow$ $(b, a) \in R^{-1}$ and $(c, b) \in S^{-1}$

$\Rightarrow$ $(c, a) \in R^{-1}oS^{-1}$.

Thus, $(SoR)^{-1} \subseteq R^{-1}oS^{-1}$.

Similarly, $R^{-1}oS^{-1} \subseteq (SoR)^{-1}$.

Hence $(SoR)^{-1} = R^{-1}oS^{-1}$.

Hence $(SoR)^{-1} = R^{-1}S^{-1}$.

Universal Relation: Let A be any set and R be the set A = A then R is called the universal relation on A.

Void Reation in a Set: Every subset of A × A is a relation on A. Since ϕ is also a subset of A × A. Therefore the null set ϕ is also a relation on A the relation is called the empty relation or void relation on A.

BINARY RELATION

Let A be a non-empty set. Then, a subset of A × A is called a binary relation on A.

The set $I_A = \{(a, a): a \in A\}$ is know as a **identity relation** on A.

VARIOUS TYPES OF RELATIONS ON A SET

Let R be a binary relation on a set A. Then R is said to be

(i) **Reflexive,** if a R a $\forall$ a $\in$ A, *i.e.,* $(a, a) \in R\ \forall\ a \in A$;

(ii) **Symmetric,** if a R b $\Rightarrow$ bRa *i.e.,* $(a, b) \in R \Rightarrow (b, a) \in R$;

(iii) **Transitive,** if a R b, b R c $\Rightarrow$ a R c

i.e., $(a, b) \in R, (b, c) \in R \Rightarrow (a, c) \in R$;

(iv) **Antisymmetric,** if a R b, b R a $\Rightarrow$ a = b.

i.e., $(a, b) \in R, (b, a) \in R \Rightarrow a = b$.

Equivalence Relation: *A relation which is simultaneously reflexive, symmetric and transitive is known as an equivalence relation.*

Partial Order Relation: *A relation which is reflexive, antisymmetric and transitive is known as a partial order relation.*

Order Relation: *A relation r on A is said to be an order relation, if*

(i) a, b $\in$ A and a $\neq$ b $\Rightarrow$ a R b or b R a,

(ii) a R b $\Rightarrow$ a $\neq$ b,

(iii) a R b and b R c $\Rightarrow$ a R c.

Example 1:

On a set A = {1, 2, 3}, it is easy to verify that the relation

(i) R_1 = {(1, 1), (2, 2), (3, 3), (1, 2) (2, 1)}
is reflexive, symmetric and transitive.

(ii) R_2 = {(1, 1), (2, 2), (1, 2), (2, 1)}
is symmetric and transitive but not reflexive.

(iii) R_3 = {(1, 1), (2, 2), (3, 3), (1, 2)}
is reflexive and transitive, but not symmetric.

(iv) R_4 = {(1, 1), (2, 2), (3, 3), (1, 2) (2, 1), (2, 3), (3, 2)}
is reflexive but neither symmetric not transitive.

(v) R_5 = {(1, 1), (2, 2), (3, 3), (1, 2), (2, 3)}
is reflexive but neither symmetric nor transitive.

(vi) R_6 = {(1, 2), (1, 3), (3, 2)}
is transitive but neither reflexive nor symmetric.

(vii) R_7 = {(1, 1), (1, 2), (2, 3)}
satisfies none of the properties of reflexivity, symmetry and transitivity

Example 2:

Let m be an arbitrary but a fixed positive integer. For any two integers a and b, define a relation,

'Congruence modulo m', by

a $\equiv$ b (mod m) $\Leftrightarrow$ (a – b) is divisible by m.

Then, the relation R = {(a, b): a, b $\in$ Z and a $\equiv$ b (mod m)} is

(i) Reflexive, since a – a = 0 is divisible by m
$\Rightarrow$ a $\equiv$ a (mod m) $\forall$ a $\in$ Z.

(ii) Symmetric, since a – b is divisible by m
$\Rightarrow$ (b – a) is divisible by m.

(iii) Transitive, since (a – b) is divisible by m and (b – c) is divisible by m
$\Rightarrow$ (a – b + b – c) = (a – c) is divisible by m.

So, r is an equivalence relation on Z.

Example 3:

On the set Z of all integers, the relation

$$R = \{a, b): a, b \in Z \text{ and } a < b\}$$

is neither reflexive nor symmetric, since

$(a, a) \not\in R$ for any $a \in Z$ $[\because a < a]$

and $(a, b) \in R \neq (b, a) \in R$ $[\because a < b \neq b < a]$.

The above relation R is transitive, since

$(a, b) \in R, (b, c) \in R$

$\Rightarrow$ $a < b$ and $b < c \Rightarrow a < c$

$\Rightarrow$ $(a, c) \in R$.

Similarly, the relation '>' on integers is neither reflexive, nor symmetric, but it is transitive However, the relation '≤' on integers is reflexive and transitive but it is not symmetric.

This relation is antisymmetric, since $a \leq$ and $b \leq a \Leftrightarrow a = b$.

Thus, it is a partial order relation on Z.

Note: A relation which is not symmetric, is not necessarily antisymmetric. For example, the relation

$R = \{(a, b): a, b \in Z, a/b\}$, where a/b denotes 'a is a divisor of b', is neither symmetric nor antisymmetric.

Example 4:

It is easy to verify that each one of the relations:

(i) *Parallelism,* on the set of all straight lines in a plane:

(ii) *Congruence,* on the set of all triangles in a plane:

and (iii) *Similarity', on the set of all triangles in a plane is an equivalence relation.*

Theorem:

A relation R on a set A is

(i) **Reflexive** $\Leftrightarrow I_A \subseteq R$, where $I_A = \{(a, a): a \in A\}$;

(ii) **Symmetric** $\Leftrightarrow R^{-1} = R$;

(iii) **Transitive** $\Leftrightarrow RoR \subseteq R$.

Proof:

(i) by definition of reflexivity, we have

R is reflexive $\Leftrightarrow a R a \; \forall \; a \in A$

$\Leftrightarrow$ (a, a) $\in$ R a $\forall$ a $\in$ A

$\Leftrightarrow I_A \subseteq R$.

(ii) Let R be symmetric. Then,

(a, b) $\in$ R $\Leftrightarrow$ (b, a) $\in$ R [$\because$ R is symmetric]

$\Leftrightarrow$ (a, b) $\in R^{-1}$

Thus $R = R^{-1}$,

So, when R is symmetric, then $R = R^{-1}$.

Again, let $R^{-1} = R$. Then

(a, b) $\in$ R $\Leftrightarrow$ (b, a) $\in R^{-1}$

$\Leftrightarrow$ (b, a) $\in$ R [$\because R^{-1} = R$]

This show that whenever $R^{-1} = R$, then R is symmetric.

Hence R is symmetric $\Leftrightarrow R^{-1} = R$.

(iii) Let R be transitive and let (a, c) be an arbitrary element of RoR. Then,

(a, c) $\in$ RoR

$\Rightarrow$ $\exists$ b $\in$ A such that (a, b) $\in$ R and (b, c) $\in$ R

$\Rightarrow$ (a, c) $\in$ R [by transitivity of R]

Thus, RoR $\subseteq$ R.

So, whenever R is transitive, then RoR $\subseteq$ R.

Again, let RoR $\subseteq$ R and let (a, b) $\in$ R, (b, c) $\in$ R, then

(a, b) $\in$ R, (b, c) $\in$ R $\Rightarrow$ (a, c) $\in$ RoR

$\Rightarrow$ (a, c) $\in$ R [$\because$ RoR $\subseteq$ R]

This shows that R is transitive.

Thus, whenever RoR $\subseteq$ R, then R is transitive.

Hence, R is transitive $\Leftrightarrow$ RoR = R.

EQUIVALENCE CLASSES

Let R be an equivalence relation on a set A. Let a $\in$ A. Then, the set of all those elements of A, which are related to a is called an equivalence class determined by a and it is denoted by [a] or $\bar{a}$.

Thus, [a] = {b $\in$ A: (a, b, $\in$ R} = {b $\in$ A: aRb].

Quotient Set

Let S be a non-empty set and R be an equivalance relation defined in S. The set of mutually disjoint equivalance classes in which S is partitioned relative to the equivalance relation R is said to be quotient set of S for the equivalance relation R, and is denoted by S/R or By S.

PROPERTIES OF EQUIVALENCE CLASSES

Theorem 1:

An equivalence relation R on set A, partitions it into mutually disjoint equivalence classes in such a way that any two elements from the same class are not related to each other and the elements from different classes are not related to each other.

Conversely, *every partition of a set A determines an equivalence relation on A.*

Proof:

Let R be an equivalence relation on a set A. Let a be an arbitrary element of A. Then, by reflexivity a R a $\forall$ a $\in$ A and therefore, a $\in$ [a].

Thus, every element of A belongs to some equivalence class i.e, A is the union of some equivalence classes. Moreover, any two equivalence classes are either disjoint or identical.

Thus, A is expressible as the union of some mutually disjoint equivalence classes.

Now, let x and y be any two elements from the same class [a].

Then x $\in$ [a] and y $\in$ [a]

$\Rightarrow$ x R a and y R a

$\Rightarrow$ x R a and a R y [by symmetry]

$\Rightarrow$ x R y [by transitivity].

This has that any two elements from the same class are related to each other.

Again, let [a] $\neq$ [b] and let x $\in$ [a] and y$\in$[b], and we must show that x is not related to y:

Now x $\in$ [a] and y $\in$ [b]

$\Rightarrow$ x R a and y R b.

Now, if possible, let x R y. Then

x R y, y R b $\Rightarrow$ x R b.

And x R a, x R b

$\Rightarrow$ a R x, x R b [by symmetry]

$\Rightarrow$ a R b

$\Rightarrow$ a $\in$ [b]

$\Rightarrow$ [a] = [b].

Hence it is contradiction.

Thus, it follows that any two elements from two different classes are not related to each other.

As a consequence of the above results, it follows that A is expressible as the union of mutually disjoint equivalence classes satisfying the requisite properties and therefore the theorem follows.

Conversely, let $\{A_1, A_2, A_3,...\}$ constitute the partition of A, so that A'³ are mutually disjoint and $A = A_1 \cup A_2 \cup A_3 \cup ...$ Consider the relation R on A defined by,

$$a \text{ R } b \Leftrightarrow a, b \in A_j \text{ for a fixed } j.$$

Then, R is

(i) *Reflexive,* for if a is an arbitrary element of A, then a is contained in some A, and since any two elements from the same A, are related to each other, so in particular aRa.

(ii) *Symmetric,* for, if a R b, then

a R b

$\Rightarrow$ a, b $\in A_j$ for a fixed j

$\Rightarrow$ b, a $\in A_j$ for a fixed j

$\Rightarrow$ a, R a.

(iii) *Transitive,* for, if a R b and b R c, then

a R be and b R c

$\Rightarrow$ a, b $\in A_j$ for some fixed j and b, c $\in A_j$ for some fixed j.

$\Rightarrow$ a, c $\in A_j$ for some fixed j [$\because A_j \cap A_j = \phi$ for $i \neq j$]

$\Rightarrow$ a R c.

Thus, R is an equivalence relation on A and A's are precisely the equivalence classes determined by the elements of A.

Theorem 2:

Let R be an equivalence relation on a set A and let us denote by [a], the equivalence class determined by the element a, then

(i) $a \in [a]$,

(ii) $a \in [b] \Leftrightarrow [a] = [b]$,

(iii) *The equivalence classes determined by two elements are either disjoint or identical.*

i.e., either $[a] \cap [b] = \phi$ *or* $[a] = [b]$.

Proof:

(i) The relation R being reflexive, we have

$a\ R\ a\ \forall\ a \in A$, Hence relative R being reflexive

And therefore, $a \in [a]\ \forall\ a \in A$.

(ii) Let $a \in [b]$, then $a\ R\ b$.

So, $x \in [a] \Leftrightarrow x\ R\ a$

$\Leftrightarrow x\ R\ a$ and $a\ R\ b$ (given)

$\Leftrightarrow x\ R\ b$ (by transivity)

$\Leftrightarrow x \in [b]$.

$\therefore\ = [a] = [b]$.

Thus $a \in [b] \Leftrightarrow [a] = [b]$.

Again, let $[a] = [b]$

Then from (i), $a \in [a]$ and so $a \in [b]$ $(\because [a] = [b])$

Hence $a \in [b] \Leftrightarrow [a] = [b]$.

(iii) For any two classes, [a] and [b], if $[a] \cap [b] = \phi$, then the result follows. So , let us consider the case, when $[a] \cap [b] \neq \phi$.

Let $x \in [a] \cap [b]$.

Then $x \in [a] \cap [b]$.

$\Rightarrow x \in [a]$ and $x \in [b]$

$\Rightarrow x\ R\ a$ and $x\ R\ b$

$\Rightarrow a\ R\ x$ and $x\ R\ b$ [by symmetry]

$\Rightarrow a\ R\ b$ (by transitivity)

$\Rightarrow a \in [b] \Rightarrow a\ [a] = [b]$.

Thus, either $[a] \cap [b] = \phi$ or $[a] = [b]$.

Example:

On the set Z of all integers, consider the relation

$$R = \{(a, b): a \equiv b \pmod 3\}.$$

This has earlier been shown that the above relation is an equivalence relation.

The equivalence classes determined by the integers 1, 2, and 3 are the set consisting of integers related to 1, 2, and 3 respectively.

Now $[1] = \{..., -8, -5, -2, 1, 4, 7, 10,...,\}$

$[2] = \{..., -7, -4, -1, 2, 5, 8, 11,...,\}$

and $[3] = \{..., -6, -3, 0, 3, 6, 9, 12, 15,...,\}$.

It may be noted here that, $Z = [1] \cup [2] \cup [3]$, where [1], [2] and [3] are mutually disjoint equivalence classes.

In the above case, we say that the sets [1], [2] [3] and constitute the partition of Z.

Partitions: Let S be a non exmpty set A set P = {A, B, C...} of non empty subset of S will be called a partition of S if:

(i) $A \cup B \cup C \cup ... = S$ *i.e.* The set S is the union of the sets in P and

(ii) The intersection of every pair of definite subset of $S \in P$ is the null set *i.e.*, if A and $B \in P$ then either $A = B$ or $A \cap B = \phi$.

BINARY OPERATION OR BINARY COMPOSITION ON A SET

Let G be a non empty set. Then an operation $*$ on a non-empty set G is said to be binary, if

$$a \in G,\ b \in G \Rightarrow a * b \in G\ \forall\ b \in G.$$

The above property is called the closure property and if this is satisfied, we say that G is closed under the operation

Example:

Addition operation (+) as well as multiplication operation (×) is a binary operation on the set N of all natural numbers, since

$a \in N, b \in N$

$\Rightarrow$ $a + b \in N\ \forall\ a, b \in N$

and $a \in N, b \in N$

$\Rightarrow$ $a \times b \in N\ \forall\ a, b \in N.$

Similarly each one of addition and multiplication operations, is a binary operation on each of the sets, the set Z of integers, the set Q of rational numbers, the set R of all real numbers and the set C of all complex numbers.

However, the set P of all irrationals, is not closed for multiplication, since $\sqrt{2} \in P$, $\sqrt{2} \in P$, but $\sqrt{2} \times \sqrt{2} = 2 \notin P$.

Similarly, subtraction operation (–) on N is not a binary composition and so is the case with the division operation (÷) on all the above sets.

Exponential operation $[(a, b) \rightarrow a^b]$ is a binary composition on each of the sets N, Q and R, but it is not a binary composition on the set Z of integers, since

$$2 \in Z, -2 \in Z \text{ but } 2^{-2} = \frac{1}{2^2} \notin Z.$$

SOME OTHER RESULTS ON OPERATIONS ON SETS

Theorem 1:

For any sets A and B, prove that:

(i) $A - B = A \cap B'$

(ii) $(A - B) \cup B = A \cup B.$

(iii) $(A - B) \cap B = \phi.$

Proof:

(i) Let x be an arbitrary element of (A – B).

Then, $x \in (A - B) \Rightarrow x \in A$ and $x \notin B$

$\Rightarrow x \in A$ and $x \in B'$

$\Rightarrow x \in (A \cap B').$

$\therefore \quad (A - B) \subseteq (A \cap B')$

Similarly, $(A \cap B') \subseteq (A - B)$

Hence $(A - B) = (A \cap B').$

(ii) Let x be an arbitrary element of $(A - B) \cup B$

Then $x \in (A - B) \cup B$

$\Rightarrow x \in (A - B) \Rightarrow x \in B$

$\Rightarrow (x \in A \;\&\; x \notin B) \Rightarrow (x \in B)$

$\Rightarrow (x \in A \Rightarrow x \in B)$ and $(x \notin B \Rightarrow x \in B)$ [$\because$ 'or' distributes '&']

$\Rightarrow x \in (A \cup B)$

$\therefore \quad (A - B) \cup B \subseteq (A \cup B)$ ...(i)

Again, let y be an arbitrarily element of $(A \cup B)$. Then

$y \in (A \cup B)$

$\Rightarrow (y \in A \Rightarrow y \in B)$

$\Rightarrow (y \in A \Rightarrow y \in B)$ and $(y \notin B \Rightarrow y \in B)$ **[Note]**

$\Rightarrow (y \in A \;\&\; y \notin B) \Rightarrow (y \in B)$ ['or' distributes '&']

$\Rightarrow y \in (A - B) \Rightarrow (y \in B)$

$\Rightarrow y \in (A - B) \cup B$ ...(ii)

Hence from (i) and (ii), we have $(A - B) \cup B = (A \cup B)$.

(iii) If possible, let $(A - B) \cap B \neq \phi$ and let $x \in (A - B) \cap B$.

Then, $x \in (A - B) \cap B$

$\Rightarrow x \in (A - B)$ and $x \in B$

$\Rightarrow (x \in A \;\&\; x \notin B)$ and $(x \in B)$

$\Rightarrow x \in A$ and $(x \notin A \;\&\; x \in B)$

But, $x \notin B$ and $x \in B$ both can never hold simultaneously.

Thus, we arrive at a contradiction.

Since the contradiction arises by assuming that $(A - B) \cap B \neq \phi$, hence $(A - B)\square \cap B \neq \phi$.

Theorem 2:

Using various laws on operations on sets, prove the following:

(i) $A \cap (B \Delta C) = (A \cap B) \Delta (A \cap C)$

(ii) $A - (A - B) = A \cap B$

(iii) $(A \cup B) - (A \cap B) = (A - B) \cap (B - A)$.

Proof:

(i) We have $A \cap (B \Delta C)$

$= A \cap [(B - C) \cup (C - B)]$

$= [A \cap (B - C)] \cup [A \cap (C - B)]$ (distributive law)

$= [(A \cap B) - (A \cap C)] \cap (A \cap C) - (A \cap B)]$

$= (A \cap B) \Delta (A \cap C)$.

(ii) We have

$A - (A - B) = A - (A \cap B')$ [$\because A - B = A \cap B'$]

$= A \cap (A \cap B')'$

$= A \cap (A' \cup B')$ [by De-Morgan's law]

$= (A \cap A') \cup (A \cap B)$

$= \phi \cup (A \cap B) = A \cap B$.

(iii) $(A \cup B) - (A \cap B)$

$= (A \cup B) \cap (A \cap B)'$

$= (A \cap B) (A' \cup B')$ [De-Morgan's law]

$= [A \cap (A' \cup B')' \cup [\ B \cap (A' \cup B')]$ (distributive law)

$= [(A \cap A) \cup (A \cap B')] \cup [(B \cap A') \cup (B \cap B')]$

(distributive law)

$= [\phi \cup (A \cap B')] \cup [(B \cup A') \cup \phi]$

$= (A \cap B') \cup (B \cap A')$

$= (A - B) \cup (B - A)$.

Example:

Let A = N and for each $\lambda \in A$.

Let $A_\lambda = \{1, 1/2, 1/3\}$

Let $U = \{x \in R, 0 \leq x \leq 1\}$

Find $U\{A_\lambda: \lambda \in A$.

Solution:

Here $\lambda = 1, 2, 3, 4, \ldots$, so that we have

$A_1 = \{1\}, A_2 = \{1, 1/2\}. A_3 = \{1, 1/2, 1/3\}$ etc.

it is clear that

$U\{A_\lambda: \lambda \in A\} = \{1, 1/2, 1/3, \ldots 1/\lambda, 1/\lambda+1\}$

ORDERED PAIR

By the definition of equality of sets, for any two objects a and b we have, $\{a, b\} = \{b, a\}$. However, if we keep in mind the order in which the elements are being listed, then *a set of two elements whose elements have been listed in a specific order is called an* **ordered pair** to be denoted by (a, b). Thus for different elements a and b we have, $(a, b) \neq (b, a)$. In general, $(a_1, b_1) = (a_2, b_2) \Leftrightarrow a_1 = a_2$ and $b_1 = b_2$.

LAWS OF BINARY COMPOSITION

A binary $*$ composition * on a non-empty set G is said

(i) **Commutative,** let

$a * b = b * a \ \forall\ a, b \in G$;

(ii) **Associative,** let

$(a * b) * c = (b * c) \ \forall\ a, b, c, \in G$;

(iii) *Also, if e is an identity element, with respect to* $*$ *in G, then an element* $a \in G$ *is said to be* **invertible,** *if there exists an element* $b \in G$ *such that,*

$$a * b = b * a = e$$

and in this case a and b are said to be the inverses of each other, written as, $a^{1} = b$ *and* $b^{1} = a$*;*

(iv) **to have an identity element,** *if there exists an element denoted by e in G such that,*

$$a * e = e * a \ \forall \ a, \in G;$$

(v) *to satisfy the* **left cancellation law,** *if*

$$a * b = a * c \Rightarrow b = c \ \forall \ a, b, c \in G;$$

(vi) *to satisfy the* **left cancellation law, if**

$$b * a = c * a \Rightarrow b = c \ \forall \ a, b, c \in G;$$

(vii) to be **right distributive** over another composition o on G. if

$$a * (boc) = (a * b) \text{ o } (a * c) \ \forall \ a, b, c \in G;$$

and (vii) to be **left distributive** over o, if

Algebraic Structures: A non-empty set G along with one or more binary compositions on G is called an algebraic structure or an *algebraic system.*

Thus, each on of the systems (N, +), (Q, ×) (R, +, ×) and [P (A), ∪, ∩)] is an algebraic structure.

Example:

Addition operation on the set R of all real numbers satisfies the closure property, associativity and commutativity; zero is the additive identity and each real number a has –a as its additive immerse.

Multiplication on R also satisfies the closure property, the associativity and the commutativity. Multiplication on R distributes addition. Also, 1 is the multiplicative identity in R and every non-zero real number a has (1/a) as its multiplicative inverse. However, 0 has no multiplicative inverse.

UNIVERSAL SET

While making a study of sets, there happens to be a set, which is a superset of each one of the sets under consideration. Such a set is known as the **Universal Set.**

Examples:

(i) If A = {1, 2, 3, 4}, B = {2, 4, 6, 8}, C = {1, 3, 5, 7} then

X = {1, 2, 3, 4, 5, 6, 7, 8} is clearly a universal set.

Power Set: *The family or collection of all subsets of a given set A, is called the* **power set** *of A and it is denoted by* P(A).

(ii) If B = {1, {2}} then, we may write B = {1, A} where A = {2}

∴ P (B) = {ϕ, {1}, {A}, {1, A}}

= {ϕ, {1}, {{2}, {1, {2}}}

(iii) If C = {{1}, {2, 3}} then we may write,

C = {A, B}, where A = {1}, B = {2, 3}

∴ P(C) = {ϕ, {A}, {B}, {A, B}}

= {ϕ, {{1}}, {{2, 3}}, {{1}, {2, 3}}}.

OPERATION ON SETS

1. **Union of Sets.** *The union of two sets A and B denoted by A ∪ B is the set constituting of all those elements, each one of which is contained either in A or in B or in both A and B.*

Thus $A \cup B = \{x: x \in A \text{ or } x \in B\}$

Consequently, $x \in A \cup B \Rightarrow x \in A \text{ or } x \in B.$

And $x \notin A \cup B \Rightarrow x \notin A \text{ and } x \notin B.$

Note: A wood about use of "or" in ordinary English when we say that something is one of the other we imply that it is not both. The mathematical "or" is quite different, at least when we are speaking in set theory. For when we say that x is in A or X is in B we mean x is in atleast one of A or B, and may be both.

Example:

(i) If $A = \{x: x \text{ is a negative integer}\}$

and $B = \{x: x \text{ is a positive integer}\}$

then $A \cup B = \{x: x \text{ is an integer}, x \neq 0\}.$

2. **Intersection of Set:** *The intersection of two sets A and B denoted by A ∩ B is the set consisting of all those elements which are common to both A and B.*

Thus $A \cap B = \{x: x \in \text{ and } x \in B\}$

Consequently, $x \in A \cap B \Rightarrow x \in A \text{ and } x \in B.$

And $x \notin A \cap B \Rightarrow x \in A \text{ or } x \notin B.$

When the sets A and B are **disjoint**, no element is common to both A and B and thus in this case, $A \cap B = \phi$.

When $A \cap B \neq \phi$, the sets A and B are said to be intersecting.

Example:

(i) Let $A = \{1, 2, 3\}$

and $B = \{2, 4, 6\}$,

then $A \cup B = (1, 2, 3, 4, 6\}$.

(ii) If $A = \{x: x = 2n + 1, n \in Z\}$

and $B = \{x: x = 2n, n \in Z\}$.

then $A \cup B = \{x: x \text{ is an odd integer}\}$

$\cup\{x: x \text{ is an even integer}\}$

$= \{x: x \text{ is an integer}\} = Z.$

Note:

(1) Each element in a set is listed once only because the repetition of elements is meaningless in a set. If A is a set of cricket players, B is a set of tennis players and some players are common to both the teams, then

$A \cup B$ = Set of all players in the two teams.

(2) From the definition it is clear that A and B are the subsets of $A \cup B$.

Symbolically, $A \subset (A \cup B)$ and $B \subset (A \cup B)$.

(3) The union of a finite number of sets $A_1, A_2, \ldots, A_n$ is denoted by $A_1 \cup A_2 \cup A_3, \ldots \cup A_n$

or by $\bigcup_{i=1}^{n} A_1$.

Example:

(i) If $A = \{x: x \text{ is a +ve integer, } x \text{ is a multiple of } 4\}$

and $B = \{x: x \text{ is a +ve integer, } x \text{ is a multiple of } 6\}$

then $A \cap B = \{x: x \text{ is a +ve integer, } x \text{ is a multiple of } 12\}$.

Examples:

(i) If $A = \{a, b, i, z\}$

$B = \{b, c, i, u, v\}$

Then $A \cap B = \{b, i\}$.

(ii) If $A = \{x: x = 2n, n \in Z\}$ and

$B = \{x: x = 3n, n \in Z\}$,

Then $A \cap B = \{x: x = 2n, n \in Z\} \cap \{x: x = 3n, n \in Z\}$

$= \{\ldots, -4, -2, 0, 2, 4, 6, \ldots\} \cap \{\ldots, -9, -6, -3, 0, 3, 6, 9, \ldots\}$

$= \{\ldots, -6, 0, 6, 12, \ldots\} = \{x: x = 6n, n \in Z\}$.

Note:

(1) If A and B are any two sets, then $A \cap B \subset A$ and $A \cap B \subset B$.

(2) The intersection of finite number of sets $A_1, A_2, \ldots, A_n$ is denoted by

$$A_1 \cap A_2 \cap \ldots \cap A_n$$

or by $\bigcap_{i=1}^{n} A_i$

3. Difference of Sets: *If A and B are two sets, then their difference denoted by A – B is defined by*

$A - B = \{x: x \in A \text{ and } x \notin B\}$

Thus $x \in A - B \Rightarrow x \in A$ and $x \notin B$.

Also the *symmetric difference* $A \Delta B$ is defined as

$$A \Delta B = (A - B) \cup (B - A).$$

Example:

If $A = \{a, b, c, d\}$ and $B = \{b, d, e, f\}$

then $A - B = \{a, c\}$ and $B - A = \{e, f\}$

and $A \Delta B = \{a, c, e, f\}$

Note that in general, $(A - B) \neq (B - A)$.

4. Disjoint Sets: *Two sets A and B are said to be disjoint sets if they have no element in common, i.e., if their intersection is a null set, i.e.,*

$$A \cap B = \phi.$$

Thus, $A \cap B = \phi \Rightarrow$ A and B are disjoint. For example, if

$$A = \{a, b, c\} \text{ and } B = \{p, q, r\}$$

then A and B are disjoint sets as there is no element common to both of them.

Example: If $A = \{2, 3, 4, 5, 6, 7\}$

and $B = \{3, 5, 7, 9, 11, 13\}$,

then $A - B = \{2, 4, 6\}$

and $B - A = \{9, 11, 13\}$.

Symmetric Difference of Two Sets: Let A and B be two sets. The symmetric difference of sets A and B is the set (A   B) ∪ (B – A) and is denoted by A Δ B.

Thus, $A \Delta B = (A - B) \cup (B - A)$

$= \{x: x \; A \cap B\}$

In following figure shaded part represents A Δ B.

For example, If $A = \{1, 2, 3\}$

$B = \{3, 4, 5)$

then $A - B = \{1, 2\}$

$B - A = \{4, 5\}$

$\therefore \; A \Delta B = \{1, 2, 4, 5\}.$

COMPLEMENT OF A SET

The complement of a set B relation to antother set A is the set of all elements which belongs to A but which do not belong to B and is denoted by A~B or by A – B, it sometime called the difference of A and B and read as "A difference B". Or,

Let A be a subset of a universal set X. Then the complement of A in X, written as X ~ A or N is defined as $A' = \{x \in X: x \notin A\}$

Thus $x \in A' \Leftrightarrow x \notin A$

Example 1:

(i) If X = {x: x is a letter in the English alphabet}

and A = {a, e, i, o, u}

then A' = {x: x is a consonant in the English alphabet}

(ii) If X = {a, b, c, d, e} and A = {b, d}

then A' = {a, c, e}.

Example 2:

Le N, the set of all natural numbers be taken as the universal set and let

A = {x: x is an even number and $x \in N$},

then A' = {x: x is an odd number, $x \in N$}.

Note:

(1) The set A and its complement A' are disjoint sets.

(2) Let A and B be two sets. The set $X = \{x: x \in A, x \notin B)$ is called the complement of the set B with respect to A (*i.e.*, A – B).

SOME RESULTS ON COMPLEMENTATION

(i) $\phi' = \{x \in X: x \notin \phi\} = X$

(ii) $X' = \{x \in X: x \notin X\} = \phi$

(iii) $(A')' = \{x \in X: x \notin A'\} = \{x \in X: x \in A\} = A$

(iv) $A \cap A' = \{x \in X: x \in A\} \cap \{x \in X: x \notin A\} = \phi$

(v) $A \cup A' = \{x \in X: x \in A\} \cup \{x \in X: x \notin A\} = X.$

LAW OF OPERATIONS

If A, B, C are any subsets of a set X, then

(a) **De-Morgan's laws**

(i) $(A \cup B)' = A' \cap B'$

(ii) $(A \cap B)' = A' \cup B'.$

(b) **Distributive laws**

(i) $A \cup (B \cap C) = (A \cup B) \cap (A \cup C)$

(ii) $A \cap (B \cup C) = (A \cap B) \cup (A \cap C)$

(c) **Associative laws**

(i) $(A \cup B) \cup C = A \cup (B \cup C)$

(ii) $(A \cap B) \cap C = A \cap (B \cap C)$

(d) **Commutative laws**

(i) $A \cup B = B \cap A$

(ii) $A \cap B = B \cap A.$

(e) **Identity laws**

(i) $A \cup \phi = A$

(ii) $A \cap X = A$

(f) **idempotent laws**

(i) $A \cup A = A$

(ii) $A \cap A = A.$

Example 1:

To prove: $A \cup B = B \cup A$.

Proof:

Let x be an arbitrary element of $A \cup B$.

Then $x \in A \cup B \Rightarrow x \in A$ or $x \in B$

$\Rightarrow x \in B$ or $x \in A$

$\Rightarrow x \in B \cup A$

$\therefore \quad A \cup B \subseteq B \cup A.$

Similarly, $B \cup A \subseteq A \cup B$.

Hence $\quad A \cup B = B \cup A$.

Example 2:

To prove: $A \cup (B \cap C) = (A \cup B) \cap (A \cup C)$.

Proof:

Let x be an arbitrary element of $A \cup (B \cap C)$.

Then $\quad x \in A \cup (B \cap C)$

$\Rightarrow x \in A$ or $x \in (B \cap C)$

$\Rightarrow x \in A$ or $(x \in B$ and $x \in C)$

$\Rightarrow (x \in A$ or $x \in B)$

and $(x \in A$ or $x \in C)$ [$\because$ 'or' distributes 'and']

$\Rightarrow x \in (A \cup B)$ and $x \in (A \cup C)$

$\Rightarrow x \in (A \cup B) \cap (A \cup C)$

$\therefore \quad A \cup (B \cap C) \subseteq (A \cup B) \cap (A \cup C)$

Similarly, $(A \cup B) \cap (A \cup C) \subseteq A \cup (B \cap C)$.

Hence $A \cup (B \cap C) = (A \cup B) \cap (A \cup C)$.

Example 3:

To prove: $(A \cap B) \cap C = A \cap (B \cap C)$.

Proof:

Let x be an arbitrary element of $(A \cap B) \cap C$.

Then $x \in (A \cap B) \cap C \Rightarrow x \in (A \cap B)$ and $x \in C$

$\Rightarrow (x \in A$ and $x \in B)$ and $x \in C$

$\Rightarrow x \in A$ and $(x \in B$ and $x \in C)$

$\Rightarrow x \in A$ and $x \in (B \cap C)$

$\Rightarrow x \in A \cap (B \cap C)$

$\therefore \quad (A \cap B) \cap C \subseteq A \cap (B \cap C)$

Similarly, $\quad A \cap (B \cap C) \subseteq (A \cap B) \cap C$.

Hence $\quad (A \cap B) \cap C = A \cap (B \cap C)$.

Example 4:

To prove: $(A \cap B)' = A' \cup B.$

Proof:

Let x be an arbitrary element of $(A \cap B)'$.

Then we have

$$x \in (A \cap B)' \Rightarrow x \notin (A \cap B)$$
$$\Rightarrow x \notin A \text{ or } x \notin B$$
$$\Rightarrow x \in A' \text{ or } x \in B'$$
$$\Rightarrow x \in (A' \cup B')$$

Thus $(A \cap B)' \subseteq (A' \cup B')$

Similarly, $(A' \cup B') \subseteq (A \cap B)'$

Hence $(A \cap B)' = (A' \cup B')$

EMPTY SET

A set consisting of no element at all is called an ***empty*** or a ***void*** or a ***null*** set and it is denoted by ϕ.

In Roaster method, it is denoted by { }.

Examples:

Each one of the sets given below is an empty set.

(i) {x: x is a natural number, $2 < x < 3$}

(ii) {x: x is a real number, $x^2 < 0$}

(iii) {x: $x \neq x$}.

Singleton Set: *A set consisting of a single point x is called a* **Singleton Set,** *to be denoted by {x}.* Thus {x: $x + 5 = 5$} = {0} is a singleton set, whose only member is 0.

FINITE AND INFINITE SETS

A set is said to be finite if it consists of a specific number of different element i.e., counting process can come to an end otherwise a set is said to be infinite.

Example 1:

Each one of the sets given below is a finite set.

(i) Set of all persons on earth;

(ii) {x: x is a natural number, x < 5 crores}

(iii) Set of all rivers in India.

Example 2:

Each one of the sets given below is an infinite set.

(i) Set of all points on the arc of a circle.

(ii) {x: x is a rational number, 0 < x < 1};

(iii) {x: x is multiple of 2, x is an integer};

(iv) Set of all concentric circles;

EQUAL SETS

Two sets A and B are said to be equal, written as $A = B$, if every element of A is in B and every element of B is in A, if $A \subset B$ and $B \subset A \Rightarrow A = B$. What is meant by the equality of two sets? For us this is always mean that they contain the same elements that is every element which is in one is in other, and vice-versa.

Notes:

(i) The repetition of elements in a set is meaningless.

Thus {x: x is a letter in 'follow'} = {x: x is a letter in 'wolf' *i.e.*,

$\{f, o, l, l, o, w\} = \{w, o, l, f\}$.

(ii) ϕ, {0} and 0 are all different, since ϕ is a set containing no element at all, {0} is a set containing one element, namely 0 and 0 is a number, not a set.

(iii) The elements of a set may be listed in any order. Thus {p, a, t} = {a, t, p} {t, p, a} etc.

SOLVED EXAMPLES

Example 1:

Prove that the operation $*$ on the set Z of all integers defined by $a * b = a + b + 1 \ \forall a, b \in Z$

satisfies the closure property the associativity and the commutativity.

Find the identity element. What is the inverse of an integer a?

Solution:

For all integers a and b, $a \in b \in I \Rightarrow a + b + 1 \in I$ and therefore $a * b$ is clearly an integer. So, closure property is satisfied.

Also, $(a * b) * c = (a + b) * c \Rightarrow (a + b + 1) * c$

$\Rightarrow \quad (a + b + 1) + c + 1 \Rightarrow a + b + c + 2$

and $\quad a * (b * c) = a * (b + c + 1)$

$= a + (b + c + 1) + 1 = (a + b + c) + 2$

$\therefore \quad (a * b) * c = a * (b * c)$ " a, b, c $\in$ Z.

Thus, the associative law is satisfied.

Similarly, commutative law can be established.

Now, if e is the identity element in Z for $*$, then

$$a * e = a \Rightarrow a + e + 1 = a \Rightarrow e = -1.$$

So, -1 is the identity for $*$ in Z.

Also, $a * b = -1 \Rightarrow a + b + 1 = -1 \Rightarrow b - (2 + a)$.

So, the inverse of a is $-(2 + a)$.

Example 2:

In a certain examination, the candidates can offer papers in English or Hindi or both the subjects. The number of candidates who appeared in the examination is 1000 of whom 650 appeared in English and 200 appeared in both English and Hindi. Find,

(i) The number of candidates who offered paper in Hindi.

(ii) The number of candidates who offered paper in English only.

(iii) The number of candidates who offered paper in Hindi only.

Solution:

Let A = The set of candidates who offered paper in English

B = The set of candidates who offered paper in Hindi.

Therefore, $n(A \cup B) = 1000$, $n(A) = 650$, $n(A \cap B) = 200$.

(i) We have

$$n(A \cup B) = n(A) + n(B) - n(A \cap B)$$

$\Rightarrow \quad 1000 = 650 + n(B) - 200$

$\Rightarrow \quad n(B) = 550$

$\therefore$ Number of candidates who offered paper in Hindi = 550.

(ii) Now, the set of candidates who offered paper in English only is

$$A - B = \{\text{candidate: candidate} \in A \text{ and candidate} \notin B\}$$

Now, $n(A - B) = n(A) - n(A \cap B)$

$= 650 - 200 = 450.$

Number of candidates who offered paper in English only $= 450.$

(iii) Similarly, the set of candidates who offered paper in Hindi only is

$B - A.$

Now, $n(B - A) = n(B) - n(A \cap B)$

$= 550 - 200 = 350.$

Example 3:

If $a \in N$ such that $aN = \{ax: x \in N\}$. Describe the set $3N \cap 7N$.

Solution:

We have $aN = \{ax ; x \in N\}$

$\therefore \quad 3N = [3x: x \in N\} = \{3, 6, 9, 12, \ldots,\}$

and $\quad 7N = [7x: x \in N\} = \{7, 14, 21, 28, \ldots,\}$

Hence $3N \cap 7N = \{21, 42, \ldots\}$

$= \{21x: x \in N\} = 21N.$

Example 4:

For any natural number a, we define $aN = \{ax: x \in N\}$. If $b, c, d \in N$ such that $bN \cap cN = dN$, then prove that d is the l.c.m. of b and c.

Solution:

We have

$bN = \{bx: x \in N\}$ = The set of positive integral multiples of b.

$cN = \{cx: x \in N\}$ = The set of positive integral multiples of c.

$\therefore$ $bN \cap cN$ = The set of positive integral multiples of b and c both.

$\Rightarrow bN \cap cN = \{kx: x \in N\}$, where k is the *l*.c.m. of b and c.

Hence, $\quad d$ = *l*.c.m of b and c.

Example 5:

Two finite sets have m and n elements. The total number of subsets of the first set is 56 more than the total number of subsets of the second set. Find the values of m and n.

Solution:

Let A and B be two sets having m and n elements respectively.

Then,

Number of subsets of $A = 2^m$,

Number of subsets of $B = 2^n$,

It is given that $2^m - 2^n = 56$

$\Rightarrow$ $2^n(2^{m-n} - 1) = 2^3(2^3 - 1)$

$\Rightarrow$ $n = 3$ and $m - n = 3$

$\Rightarrow$ $n = 3$ and $m = 6$.

Example 6:

The necessary and sufficient condition for a set Y to be a subset of X is that $X \cup Y = X$.

Solution:

Let $Y \subset X$, then

$$x \in Y \Rightarrow x \in X \quad \text{...(i)}$$

Now, $x \in X \cup Y \Rightarrow x \in X$ or $x \in Y$

$\Rightarrow x \in X$

$\therefore X \cup Y \subset X$ ...(ii)

Also, we know that

$$X \subset X \cup Y \quad \text{...(iii)}$$

(ii) and (iii) $\Rightarrow X \cup Y = X$.

Conversely,

if $X \cup Y = X$, we have to prove that $Y \subset X$.

Now, $X \cup Y = X \Rightarrow X \cup Y \subset X$

and $X \subset X \cup Y$

$\Rightarrow$ $X \cup Y \subset X$

$\Rightarrow$ $X \subset X$ and $Y \subset X$

$\Rightarrow$ $Y \subset X$.

Example 7:

Prove that $(A')' = A$.

Solution:

We have to prove $(A')' \subset A$ and $A \subset (A')'$.

Let $x \in (A')'$ then

$x \in (A')' \Leftrightarrow x \notin A'$

$\Leftrightarrow x \in A$

Thus $x \in (A')' \Leftrightarrow x \in A$

$\Rightarrow$ $(A')' \subset A$ and $(A')' \supset A$

$\Rightarrow$ $(A')' = A$.

Example 8:

Let A, B, C be three sets then

$$(A - B) \cap (A - C) = A - (B \cup C)$$

Solution:

Let $x \in (A - B) \cap (A - C)$ then

$$\begin{aligned} x \in (A - B) \cap (A - C) &\Leftrightarrow x \in (A - B) \text{ and } x \in (A - C) \\ &\Leftrightarrow x \in A, x \notin B \text{ and } x \in A, x \notin C \\ &\Leftrightarrow x \in A, x \notin B \text{ and } x \notin C \\ &\Leftrightarrow x \in A, x \notin B \cup C \\ &\Leftrightarrow x \in A - (B \cup C) \end{aligned}$$

Which implies that $(A - B) \cap (A - C) = A - (B \cup C)$

Example 9:

If A, B and C are any three sets, then prove that:

(a) $A \times (B \cap C) = (A \times B) \cap (A \times C)$,

(b) $A \times (B \cup C) = (A \times B) \cup (A \times C)$.

Solution:

(a) We shall show that

$$A \times (B \cap C) \subset (A \times B) \cap (A \times C)$$

and $$(A \times B) \cap (A \times C) \subset A \times (B \cap C)$$

Firstly, let $(x, y) \in A \times (B \cap C)$, then

$$\begin{aligned} & (x, y) \in A \times (B \cap C) \\ \Rightarrow\ & x \in A \text{ and } y \in (B \cap C) \\ \Rightarrow\ & x \in A \text{ and } [y \in B \text{ and } y \in C] \\ \Rightarrow\ & [x \in A \text{ and } y \in B] \text{ and } [x \in A \text{ and } y \in C] \\ \Rightarrow\ & (x, y) \in (A \times B) \text{ and } (x, y) \in (A \times C) \\ \Rightarrow\ & (x, y) \in (A \times B) \cap (A \times C) \end{aligned}$$

Thus, $A \times (B \cap C) \subset (A \times B) \cap (A \times C)$ ...(1)

Again, let $(x, y) \in (A \times B) \cap (A \times C)$, then

$$\begin{aligned} & (x, y) \in (A \times B) \cap (A \times C) \\ \Rightarrow\ & (x, y) \in A \times B \text{ and } (x, y) \in (A \times C) \\ \Rightarrow\ & (x \in A, y \in B) \text{ and } (y \in A, y \in C) \end{aligned}$$

$\Rightarrow$ $x \in A, y \in B \cap C$

$\Rightarrow$ $(x, y) \in A \times B \cap C$

$\therefore$ $(A \times B) \cap (A \times C) \subset A (B \cap C)$ ...(2)

From (1) and (2)

$$A \times (B \cap C) = (A \times B) \cap (A \times C)$$

(b) Firstly, let $(x, y) \in A \times (B \cup C)$, then

$(x, y) \in A \times (B \cup C)$

$\Rightarrow$ $x \in A, y \in (B \cup C)$

$\Rightarrow$ $x \in A$, [$y \in B$ or $y \in C$]

$\Rightarrow$ [$x \in A, y \in B$] or [$x \in A, y \in C$]

$\Rightarrow$ $(x, y) \in A \times B$ or $(x, y) \in A \times C$

$\Rightarrow$ $(x, y) \in (A \times B) \cup (A \times C)$.

Since, $(x, y) \in A \times (B \cup C)$

$\Rightarrow$ $(x, y) \in (A \times B) \cup (A \times C)$

$\therefore$ $A \times (B \cup C) \subset (A \times B) \cup (A \times C)$ ...(1)

Similarly, it can be proved that

$(x, y) \in (A \times B) \cup (A \times C)$

$\Rightarrow$ $(x, y) \in A \times (B \cup C)$,

$\therefore$ $(A \times B) \cup (A \times C) \subset A \times (B \cup C)$ ...(2)

From (1) and (2)

$$A \times (B \cup C) = (A \times B) \cup (A \times C)$$

Example 10:

If $A \subset B$ show that $A \times A \subset (A \times B) \cap (B \times A)$.

Solution:

Let $(x, y) \in A \times A$, then

$(x, y) \in A \times A$,

$\Rightarrow$ $x \in A, y \in A$

Since $A \subset B$ is given,

$x \in A \Rightarrow y \in B$.

$\therefore$ $(x, y) \in A \times A$

$\Rightarrow$ $x \in A, y \in B$

$\Rightarrow$ $(x, y) \in (A \times B)$ ...(1)

Also, $(x, y) \in A \times A$

$\Rightarrow \quad x \in A, y \in A$

$\therefore \quad x \in B, y \in A$ $\quad [\because A \subset B]$

$\Rightarrow \quad (x, y) \in B \times A$ ...(2)

Hence, from (1) and (2),

$(x, y) \in A \times A$

$\Rightarrow \quad (x, y) \in (A \times B)$ and $(x, y) \in B \times A$

$\Rightarrow \quad (x, y) \in (A \times B) \cap (B \times A)$

$\therefore \quad A \times A \subset (A \times B) \cap (B \times A).$

Example 11:

If $A \subset B$, prove that $A \times C \subset B \times C$ for any set C.

Solution:

Let (x, y) be any arbitrary element of $A \times C$. Then,

$$(x, y) \in A \times C$$

$\Rightarrow \quad x \in A$ and $y \in C$

$\Rightarrow \quad x \in B$ and $y \in C$ $\quad [\because A \subset B]$

$\Rightarrow \quad (x, y) \in B \times C$

$\therefore \quad A \times C \subset B \times C.$

Example 12:

Prove, using mathematical induction that $10^{2n\ 1} + 1$ is divisible by 11 for all $n \in N$.

Solution:

Let P(n) be the statement.

P(n): "$10^{2n-1} + 1$ is divisible by 11."

When $n = 1$, we get $10^{2n-1} + 1 = 10 + 1 = 11$,

Which is divisible by 11. Hence, P(1) is true.

Let the result be true for P(r), *i.e.*, $10^{2r-1} + 1$ is divisible by 11. Therefore,

$10^{2r-1} + 1 = 11k, k \in N.$

or $\quad 10^{2r-1} = 11k - 1.$

We are now to prove that P(r + 1) is true.

Now, $\quad = P(r + 1): 10^{(2r+1)-1} - 1 = 10^{2r+1} + 1$

$= 10^{2r-1}.10^2 + 1 + (11k-1)\ 100 + 1 = 11(100k - 9)$

= multiple of 11.

Hence, 10^{2r+1} is divisible by 11. That is, P(r + 1) is true. By the principle of mathematical induction P(n) is true for all $n \in N$.

Example 13:

Using mathematical induction prove that $(1 + x)^n > 1 + nx$, for $n \geq 2$ and $x > -1, (\neq 0)$.

Solution:

Let the given result be denoted by P(n).

When n = 1, we get

$(1 + x) > 1 + x$

Which is not true (this is given in the problem).

When n = 2, we get

$(1 + x)^2 > 1 + 2x + x^2 > 1 + 2x$

Which is true. Therefore, P(2) is true.

Let the result P(r) be true, *i.e.*,

$(1 + x)^r > 1 + rx$ ...(1)

We are to prove that P (r + 1) is true,

i.e., $(1 + x)^{r+1} > 1 + (r + 1)\ x$

Now $(1 + x)^{r+1} = (1 + x)^r\ (1 + x)$

$> (1 + rx)\ (1 + x)$ (using (1))

$= 1 + rx + x + rx^2$

$= 1 + (r + 1)\ x + rx^2$

$> 1 + (r + 1)\ x$

Therefore P (r + 1) is also true. Hence, P (n) is true for all $n \geq 2, n \in N$.

Example 14:

By mathematical induction, prove that $7^{2n} + (2^{3n-3})3^{n-1}$ is divisible by 25, $n \in N$.

Solution:

Let the statement P(n) be defined as

P(n) = "$7^{2n} + (2^{3n-3})\ 3^{n-1}$ is divisible by 25"

When n = 1, we get

$P(1) = 7^2 + 1\ (1) = 50$

Which is divisible by 25,

Let the result P(r) be true.

That is $7^{2r} + (2^{3r-3})\ 3^{r-1}$ is divisible by 25

Let $7^{2r} + (2^{3r-3})\ 3^{r-1} = 25\ k,\ k \in N$ ....(1)

$$\begin{aligned}
\text{Now}\quad P(r+1) &= 7^{2r+2} + (2^{3r})\ 3^{r} \\
&= 7^{2r}\ (49) + (2^{3r-3}\ .\ 2^{3})\ 3^{r-1} : 3 \\
&= 49\ (7^{2r}) + 24\ (2^{3r-3})\ 3^{r-1} \\
&= (50 - 1)\ (7^{2r}) + (25 - 1)\ (2^{3r-3})3^{r-1} \\
&= 50\ (7^{2r}) + 25\ (2^{3r-3})\ 3^{r-1} - [7^{2r} + (2^{3r-3})\ 3^{r-1}] \\
&= 25\ [2\ (7^{2r}) + (2^{3r-3})\ 3^{r-1}] - 25\ k \qquad \text{(from (1))} \\
&= 25\ [2\ (7^{2r}) + (2^{3r-3})\ 3^{r-4} - k] \\
&= \text{divisible by 25.}
\end{aligned}$$

Therefore, P (r + 1) is also true. By mathematical induction, P(n) is true for all $n \in N$.

Example 15:

Using mathematical induction prove that for every integer $n \geq 1$, $\left(3^{2^n} - 1\right)$ is divisible by 2^{n+2} but not by 2^{n+3}.

Solution:

Let P(n) be the statement P(n): $3^{2^n} - 1$ is divisible by 2^{n+2} but not by 2^{n+3}. For n = 1, we have $P(1) = 3^2 - 1 = 8 = 2^3$, which is divisible by 2^3 but not by 2^4.

Hence P(1) is true. Let the statement P(r) be true. That is

P (r): $3^{2^r} - 1$ is divisible by 2^{r+2} but not by 2^{r+3}

Therefore, we can write

P (r): $3^{2^r} - 1 = k\ .\ 2^{r+2}$

where k is odd (if k is even, then $k.2^{r+2}$ is divisible by 2^{r+3}).

Hence $3^{2^r} = 1 + k\ 2^{r+2}$, k odd.

$$\begin{aligned}
\text{Now,}\quad P\ (r+1):\ 3^{2^{r+1}} - 1 &= 3^{(2^r.2)} - 1 \\
&= \left(3^{2^r}\right)^2 - 1 = (1 + k\ .\ 2^{r+2})^2 - 1 \\
&= k^2\ .\ 2^{2(r+2)} + 2k\ .\ 2\ 2^{r+2} + 1 - 1 \\
&= k^2\ .\ 2^{2r+4} + k\ .\ 2^{r+3} = k^2\ .\ 2^{r+3}\ .\ 2^{r+1} + k\ .\ 2^{r+3}
\end{aligned}$$

$$= k \,.\, 2^{2r+3} [1 + k \,.\, 2^{r+1}]$$

Which is divisible by 2^{r+3}, but not by 2^{r+4} as k is odd. Hence P(r + 1) is true. By mathematical induction, P(n) is true fro all $n \in N$.

Example 16:

Let $p \geq 3$ be an integer and α, β be the roots of $x^2 - (p + 1) x + 1 = 0$. Using mathematical induction show that $\alpha^n + \beta^n$ is an integer.

Solution:

Since α, β are the roots of the equation $x^2 - (p + 1) x + 1 = 0$, we have

$\alpha + \beta = p + 1$, and $\alpha \beta = 1$ ...(1)

Let $P(n) = \alpha^n + \beta^n$. We are to prove that P(n) is an integer.

For n = 1, we get

$P(1) = \alpha + \beta = p + 1$.

Which is an integer since p is an integer.

Hence, P(1) is true. Let P(r) be true.

That is, $P(r) = \alpha^r + \beta^r$ is an integer.

Now, $P(r + 1) = \alpha^{r+1} + \beta^{r+1}$

$$= \alpha (\alpha^r) + \beta(\beta^r)$$

$$= \alpha (\alpha^r + \beta^r) - \alpha\beta^r + \beta (\alpha^r + \beta^r) - \beta \alpha^r$$

$$= (\alpha + \beta) (\alpha^r + \beta^r) - \alpha\beta(\alpha^{r-1} + \beta^{r-1}$$

$$= (p + 1) (\alpha^r + \beta^r) - (\alpha^{r-1} + \beta^{r-1}). \quad ...(2)$$

Since p is an integer, p + 1 is an integer and $\alpha^r + \beta^r$ is an integer. Therefore P (r + 1) is an integer, if $\alpha^{r-1} + \beta^{r-1}$ is an integer putting r = r – 1 in (2), we get

$$P(r) = \alpha^r + \beta^r = (p + 1) (\alpha^{r-1} + \beta^{r-1}) - (\alpha^{r-2} + \beta^{r-2}) \quad ...(3)$$

Again, since (p + 1) and $(\alpha^r + \beta^r)$ are integers, $\alpha^{r-1} + \beta^{r-1}$ is an integer, if $\alpha^{r-2} + \beta^{r-2}$ is an integer. Setting r = r–1 in (3), we find $\alpha^{r-2} + \beta^{r-2}$ is an integer this way, we get that $\alpha^2 + \beta^2$ is an integer if $\alpha + \beta$ is an integer. But $\alpha + \beta = p + 1$ is an integer. Hence, $\alpha^2 + \beta^2$ is an integer, $\alpha^3 + \beta^3$ is an integer . . . . . $\alpha^{r-1} + \beta^{r-1}$ is an integer.

Hence, by mathematical induction, $\alpha^n + \beta^n$ is an integer.

Example 17:

Prove the following

(a) $A \Delta \phi = A$,

(b) $A \Delta A = \phi$,

(b) $A \Delta B = \phi \Leftrightarrow A = B$.

Solution:

(a) $A \Delta \phi = (A - \phi) \cup (\phi - A)$

$= A \cup \phi$ $\quad (\because A - \phi = A, \phi - A = \phi)$

$= A.$

(b) $A \Delta A = (A - A) \cup (A - A)$

$= \phi \cup \phi = \phi$

(c) $A \Delta B = \phi \Rightarrow (A - B) \cup (B - A) = \phi$

$\Rightarrow A - B = \phi$ and $B - A = \phi$

$\Rightarrow A = B$

and $\quad A = B \Rightarrow A - B = \phi$ and $B - A = \phi$

$\Rightarrow (A - B) \cup (B - A) = \phi$

$\Rightarrow A \Delta B = \phi$

Hence, $\quad A \Delta B = \phi \Rightarrow A = B.$

Example 18:

If A = set of all students,

B = set of all students offering mathematics,

C = set of all women,

D = set of industrious person,

E = set of all first class students.

Write the following statement symbolically. Some industrious women students are the students of mathematics but they are not first class".

Solution:

Set of women students = $A \cap C$. Set of industrious women students

$= A \cap C \cap D$

Set of industrious women students offering mathematics

$= A \cap C \cap D \cap B.$

Set of all those industrious women students offering mathematics who are not first class = $A \cap C \cap D \cap B - E$. According to the statement there are some students of this type. Hence,

$$A \cap C \cap D \cap B - E \neq \phi.$$

Example 19:

Set A has three elements and set B has six elements. What can be the minimum number of elements in the set $A \cup B$? Find also the maximum number of elements in $A \cup B$.

Solution:

We have n(A) = 3, and n(B) = 6. Therefore, the maximum number of elements in $A \cap B$ is 3, since

$$n(A \cup B) = n(A) + n(B) - n(A \cap B)$$

the minimum number of elements in $A \cup B$ is given by

$$\text{minimum } [n(A \cup B)] = n(A) + n(B) - \text{max. } [n(A \cap B)]$$
$$= 3 + 6 - 3 = 6$$

Therefore, the minimum number of elements in $A \cup B = 6$.

The number of elements in $A \cup B$ is maximum when A and B are disjoint. In this case

$$n(A \cup B) = n(A) + n(B) = 3 + 6 = 9$$

The maximum number of elements in $A \cup B$ is 9.

EXERCISES

1. N be the set of all natural numbers. Prove that each one of the relations R_1 and R_2 on $N \times N$, defined by
   $(a, b)\ R_1\ (c, d) \Leftrightarrow a + d = b + c$;
   and $(a, b)\ R_2\ (c, d) \Leftrightarrow ad = bc$
2. Show that the set of even integers is a subset of the set of integers.
3. A = {a, b, c, d, e}, B = {a, c, e, g}
   and C = {b, e, f, g}, prove that
   $(A \cup B) \cap C = (A \cap C) \cup (B \cap C)$.
4. If A and B are any two sets, then prove that $A - B$, $A \cap B$ and $B - A$ are pairwise disjoint.
5. If A = {x: x is a factor of 20}
   B = {x: x is a multiple of 5 and x < 20}
   Find A – B and B – A.
6. If A, B are two sets, show that $A - B \neq B - A$.
7. If sets $A = \{(x, y): x^2 + y^2 = 1\}$ and $B = \{(x, y): x + y = 1$, then find $A \cap B$.
8. If A and B are any subsets, then prove that $A \cap (B - A) = \phi$.
9. For any two sets A and B, prove that $A' - B' = B - A$.
10. If $A \cup B = A \cup C$ and $A \cap B = A \cap C$, then show that B = C.

3

Logarithms of Complex Numbers

THE GENERAL EXPONENTIAL FUNCTION a^z

The general exponential function a^z is defined. If a and z are any two complex numbers, t

$$a^z = e^{z \log a}$$
$$= \exp (z \log a).$$

Since log a is a many-valued function, a^z is also a many-valued function. Thus,

$$a^z = e^{z \log a} = e^{z (\log a + 2n\pi i)}$$
$$= \exp [z (\log a + 2n\pi i)].$$

This gives the *general value of a^z*. The *principal value* of a^z is obtained by putting n = 0.

Hence the principal value of a^z

$$= e^{z \log a} = \exp (a \log a).$$

LOGARITHM OF A POSITIVE REAL NUMBER IN THE SET OF COMPLEX NUMBERS

Let x be a positive real number.

Let $x = x + i0 = r (\cos \theta + i \sin \theta)$

Then $x = r \cos \theta$, $0 = r \sin \theta$.

$\therefore$ $r = x$ and $\theta = 0$.

$\therefore$ $\log x = \log x + i0 + 2 n\pi i = \log x + 2n\pi i$.

Obviously only n = 0 gives a real value of log x.

Thus in the set of complex numbers a positive real number has an infinite number of logarithms out of which only one is real.

TO SEPARATE $(\alpha + i\beta)^{p+iq}$ INTO REAL AND IMAGINARY PARTS

By the definition of the general exponential function, we have

$(a + ib)^{p + iq} = e^{(p + iq) \log (a + ib)} = \exp [(p + iq) \log (a + ib)]$

$$= \exp\left[(p+iq)\left\{\frac{1}{2}\log\left(\alpha^2+\beta^2\right)+i\tan^{-1}\left(\frac{\beta}{\alpha}\right)+2n\pi i\right\}\right]$$

$$= \exp\left[\left\{\frac{1}{2}p\log\left(\alpha^2+\beta^2\right)-q\tan^{-1}\left(\frac{\beta}{\alpha}\right)-2nq\pi\right\}\right.$$

$$\left.+i\left\{\frac{1}{2}q\log\left(\alpha^2+\beta^2\right)+p\tan^{-1}\left(\frac{\beta}{\alpha}\right)+2np\pi\right\}\right]$$

$= \exp (A + iB)$, where $A=\frac{1}{2}p\log\left(\alpha^2+\beta^2\right)-q\left\{\tan^{-1}\left(\frac{\beta}{\alpha}\right)+2n\pi\right\}$

and $B=\frac{1}{2}q\log\left(\alpha^2+\beta^2\right)+p\left\{\tan^{-1}\left(\frac{\beta}{\alpha}\right)+2n\pi\right\}$

$= e^{A + iB} = e^A e^{iB} = e^A (\cos B + i \sin B)$.

Therefore the real part is $e^A \cos B$ and the imaginary part is $e^A \sin B$.

The principal value is obtained by putting $n = 0$.

Note: In the following examples it is sometimes found convenient to write e^x as $\exp (x)$.

LOGARITHMS OF COMPLEX NUMBERS

Now we shall extend our definition of logarithmic function to the set of complex numbers.

Definition: Let z and w be two complex numbers such that $e^z = w$. Then z is called a *logarithm of w* to the base e and we write it as $z = \log_e w$ or simply as $z = \log w$, the base e remaining understood.

Thus by definition, $z = \log w$ if and only if $w = e^z$.

Since $e^z \neq 0$ for any complex number z, therefore log w does not exist if $w = 0$.

Logarithm of a Complex Number is a Many Valued Function

Let $\log_e w = z$. Then $e^z = w$.

If n is any integer, we have

$e^{2n\pi i} = \cos 2n\pi + i \sin 2n\pi = 1 + i0 = 1$.

$\therefore e^z = e^z.1 = e^z e^{2n\pi i} = e^{z + 2n\pi i}$

which means that if log w = z, then we also have

$$\log w = z + 2n\pi i.$$

Thus logarithm of a complex number is a many-valued function.

PRINCIPAL AND GENERAL VALUES OF LOGARITHM OF A NON-ZERO COMPLEX NUMBER

Let $z = x + iy$ be a non-zero complex number. Suppose

$$\log_e z = \alpha + i\beta,$$

where α and β are real. Then

$$z = e^{\alpha + i\beta} = e^{\alpha} e^{i\beta} = e^{\alpha} (\cos \beta + i \sin \beta).$$

Since e^{α} is a positive real number, therefore

$$z = e^{\alpha} (\cos \beta + i \sin \beta)$$

is a representation of z in a modulus-argument form. We have

$$e^{\alpha} = |z| = \sqrt{(x^2 + y^2)} \text{ and } \beta = \arg z = \tan^{-1}\left(\frac{y}{x}\right).$$

The equation $e^a = |z|$ has a unique solution $\alpha = \log |z|$, the real logarithm of the positive number $|z|$.

$$\therefore \qquad \log_e z = \log |z| + i \arg z$$

$$i.e., \qquad \log_e (x + iy) = \log\sqrt{(x^2 + y^2)} + i \tan^{-1}\left(\frac{y}{x}\right).$$

Now arg z is a many-valued function and consequently log z is a many-valued function. If β_0 is the principal value of arg z *i.e.,* the value of arg z lying between $-\pi$ and π, then $\log |z| + i\beta_0$ is called *the principal value of* log z. Again if β_0 is the principal value of arg z, then $2n\pi + \beta_0$ is the general value of arg z and

$$\log |z| + i\beta_0 + 2n\pi i$$

is called *the general value of log z. Thus every non-zero complex number has infinitely many logarithms which differ from one another by an integral multiple of $2\pi i$.*

It is usual to denote the general value of $\log_e z$ by $\log_e z$, using the first letter L as the capital letter, and the principal value by $\log_e z$, using the first letter *l* as the small letter. Thus we have

$$\log_e z = \log_e z + 2n\pi i, \text{ where n is any integer.}$$

If in the general value, we put n = 0, we get the principal value.

Remember:

$\log_e (x + iy) = \log\sqrt{(x^2 + y^2)} + i \tan^{-1}\left(\frac{y}{x}\right)$,

where $\tan^{-1}\left(\frac{y}{x}\right)$ represents the principal value of arg (x + iy)

and $\log_e (x + iy) = \log\sqrt{(x^2 + y^2)} + i \tan^{-1}\left(\frac{y}{x}\right) + 2n\pi i$,

where n is any integer.

PROPERTIES OF THE LOGARITHMIC FUNCTION

If u and v are two non-zero complex numbers, then

Log (uv) = Log u + Log v ...(1)

and Log (u/v) = Log u – Log V. ...(2)

Proof: Let Log u = z_1 and Log v = z_2.

Then $e^{z_1} = u$ and $e^{z_2} = v$.

$\therefore\ e^{z_1} e^{z_2} = uv$, or $e^{z_1+z_2} = uv$.

By the definition of logarithm, we have

Log uv = $z_1 + z_2$ = Log u + Log v, which proves the result (1).

The result (2) can be proved similarly.

The equality (1) does not mean that the principal value of log uv is equal to the sum of the principal values of log u and log v. Note that the sum of the principal arguments of u and v need not equal to some value of log uv and each value of log uv is equal to some value of log u + log v.

A similar remark holds for the equality (2) also.

Remark: If z be a non-zero complex number and n be a positive integer, the equality $\log z^n = n \log z$ may not be true even for the general values.

As an illustration, we have

$\text{Log } i^3 = \text{Log } (-i) = \left(2n\pi - \frac{1}{2}\pi\right)i$,

while $3\,\text{Log } i = 3\left(2m\pi + \frac{1}{2}\pi\right)i$.

Now it is not correct to say that

$$3\left(2m\pi + \frac{1}{2}\pi\right) = 2n\pi - \frac{1}{2}\pi.$$

Now it is not correct to say that

$$3\left(2m\pi + \frac{1}{2}\pi\right) = 2n\pi - \frac{1}{2}\pi.$$

For, the left hand side may be written as

$$2(3m+1)\pi - \frac{1}{2}\pi,$$

which shows that the solution set on the left is only a subset of the solution set on the right.

Thus Log $i^3 \neq 3$ Log i.

LOGARITHM OF A NEGATIVE REAL NUMBER

Let x be a positive real number so that –x is a negative real number. We have to find log (–x).

Let $-x = -x + i0 = r(\cos\theta + i\sin\theta)$.

Then $-x = r\cos\theta$, $0 = r\sin\theta$.

$\therefore$ r = x, and $\theta = \pi$.

$\therefore \log(-x) = \log x(\cos\pi + i\sin\pi) = \log x\{\cos(2n\pi + \pi) + i\sin(2n\pi + \pi)\}$

$= \log x\, e^{(2n\pi + \pi)i} = \log x + \log e^{(2n\pi + \pi)i}$

$= \log x + (2n\pi + 1)\pi i$, which is never real.

The principal value of log (–x) is $\log x + i\pi$.

Hence, the principal value of the logarithm of a negative real number is the logarithm of the corresponding positive number added with πi.

LOGARITHMS IN THE SET OF REAL NUMBERS

We know that if a and x are two real numbers such that $e^x = a$, then x is called the logarithm of a to the base e and we write it as

$$x = \log_e a.$$

Since for all real numbers x, we have $e^x > 0$, therefore, in the case of a real variable, $\log_e a$ exists if and only if a is positive. Also if a is a positive real number, $\log_e a$ is a unique real number.

WORKING RULE TO EVALUATE Log (x + iy) *i.e.*, TO EXPRESS Log (x + iy) IN THE FORM A + iB

First we put x + iy in the modulus-argument form. So let

$x + iy = r(\cos\theta + i\sin\theta)$.

Then x = r cos θ, y = r sin θ.

$\therefore\ r\sqrt{(x^2+y^2)}$ and $\theta = \tan^{-1}\left(\frac{y}{x}\right)$.

Now (x + iy) = Log (r cos θ + ir sin θ) = Log r (cos θ + i sin θ)

$= \log [r\{\cos (2n\pi + \theta) + i \sin (2n\pi + \theta)\}] = \log r\, e^{i(2n\pi + \theta)}$

$= \log r + \log e^{i(2n\pi + \theta)} = \log r + i\,(2n\pi + \theta)$

$$= \log\sqrt{(x^2+y^2)} + i\left\{2n\pi + \tan^{-1}\left(\frac{y}{x}\right)\right\}$$

$$= \frac{1}{2}\log(x^2+y^2) + i\left\{2n\pi + \tan^{-1}\left(\frac{y}{x}\right)\right\} = A + iB,$$

where $A = \frac{1}{2}\log(x^2+y^2)$,

$$B = 2n\pi + \tan^{-1}\left(\frac{y}{x}\right).$$

If we put n = 0, we obtain the principal value of Log (x + iy) written as log (x + iy). Thus

$$\log(x+iy) = \frac{1}{2}\log(x^2+y^2) + i\tan^{-1}\left(\frac{y}{x}\right).$$

Putting –y for y on both sides, we get

$$\log(x-iy) = \frac{1}{2}\log(x^2+y^2) - i\tan^{-1}\left(\frac{y}{x}\right).$$

SOLVED EXAMPLES

Example 1:

Prove that if $(1 + i \tan \alpha)^{1 + i \tan \beta}$ *can have real values, one of them is* $(\sec \alpha)^{\sec^2 \beta}$.

Solution:

We have $(1 + i \tan \alpha)^{1 + i \tan \beta}$

= exp [(1 + i tan β) log (1 + i tan α)], taking the principal value

$$= \exp\left[(1 + i\tan\beta)\left\{\log\sqrt{(1+\tan^2\alpha)} + i\tan^{-1}(\tan\alpha)\right\}\right]$$

= exp [(1 + i tan β) (log sec α + iα)]

= exp [log sec α – α tan β + i (α + tan β log sec α)]

= exp [log sec α – α tan β] · exp [i (α + tan β log sec α)]

= exp [log sec α – α tan β] · [cos {α + tan β log sec α} + i sin {α + tan β log sec α}].

Nos if this value is real, the imaginary part is equal to zero.

∴ sin {α + tan β log sec α} = 0

or α + tan β log sec α = 0

or α = – tan β log sec α. ...(1)

Also then $(1 + i \tan \alpha)^{1 + i \tan \beta}$ = exp [log sec α – α tan β] · (cos 0 + i sin 0)

= exp [log sec α + tan β · tan β log sec α], substituting for a from (1)

= exp [(1 + $\tan^2 \beta$) log sec α] = exp [$\sec^2 \beta$ log sec α]

= exp [log (sec α)sec$^{2\ \beta}$] = $(\sec \alpha)^{\sec^2 \beta}$.

This is one of the values since $(1 + i \tan \alpha)^{1 + i \tan \beta}$ can have infinite number of values.

Example 2:

Find the general value of $\log_4 (-2)$.

Solution:

We have $\log_4 (-2) = \dfrac{\log_3(-2)}{\log_e 4}$

$$= \frac{\log_e(-2) + 2n\pi i}{\log_e 4 + 2m\pi i}, \text{ where m and n are any integers}$$

$$= \frac{\log_e\{2(\cos\pi + i \sin\pi) + 2n\pi i\}}{\log_e 4 + 2m\pi i} = \frac{\log_e(2e^{i\pi}) + 2n\pi i}{\log_e 4 + 2m\pi i}$$

$$= \frac{\log_e 2 + i\pi + 2n\pi i}{\log_e 4 + 2m\pi i}$$

$$= \frac{\log_e 2 + i(2n\pi + \pi)}{2\log_e 2 + 2m\pi i} = \frac{[\log_e 2 + i(2n\pi + \pi)][2\log_e 2 - 2m\pi i]}{(2\log_e 2 + 2m\pi i)(2\log_e 2 - 2m\pi i)}$$

$$= \frac{2(\log_e 2)^2 + 2m\pi(2n\pi + \pi) + 2i\{(2n+1)\pi \log_e 2 - m\pi \log_e 2\}}{4(\log_e 2)^2 + 4m^2\pi^2}$$

$$= \frac{(\log_e 2)^2 + (2n+1)m\pi^2}{2(\log_e 2)^2 + 2m^2\pi^2} + i\frac{(2n+1-m)\pi \log_e 2}{2(\log_e 2)^2 + 2m^2\pi^2},$$

which is required general value of $\log_4 (-2)$.

Example 3:

If sin log (i^i) = a + ib, find a and b. Hence find cos (log i^i).

Solution:

First, we have

$$i^i = \exp(i \log i) = \exp\left[i \log\left(\cos\frac{1}{2}\pi + i \sin\frac{1}{2}\pi\right)\right] = \exp[i \log e^{i\pi/2}]$$

$$= \exp\left[i\left(\frac{i\pi}{2}\right)\right] = \exp\left(-\frac{\pi}{2}\right) = e^{-\pi/2}.$$

$$\therefore \log i^i = \log^{-\pi/2} = -\frac{\pi}{2}.$$

$$\therefore \sin(\log i^i) = \sin\left(-\frac{\pi}{2}\right) = -1.$$

Hence sin (log i^i) = a + ib

$\Rightarrow -1 = a + ib$.

Equating real and imaginary parts, we get

$a = -1$, $b = 0$.

$\therefore \sin(\log i^i) = -1$.

Now $\cos(\log i^i) = \sqrt{\{1 - \sin^2(\log i^i)\}} = \sqrt{\{1-1\}} = 0.$

Example 4:

If log sin (x + iy) = a + ib, show that

(i) $\alpha = \frac{1}{2}\log\frac{1}{2}(\cosh 2y - \cos 2x)$,

(ii) $2\cos 2x = e^{2y} + e^{-2y} - 4e^{2\alpha}$,

(iii) $b = \tan^{1}(\cot x \tanh y)$,

(iv) $\cos(x - \beta) = e^{2y}\cos(x + \beta)$.

Solution:

$\because \log\sin(x + iy) = \alpha + i\beta$,

$\therefore \sin(x + iy) = e^{\alpha + i\beta} = e^{\alpha} e^{i\beta}$

or $\sin x \cos iy + \cos x \sin iy = e^{\alpha}(\cos\beta + i\sin\beta)$

or $\sin x \cosh y + i\cos x \sinh y = e^{\alpha}\cos\beta + i e^{\alpha}\sin\beta$.

Equating real and imaginary parts, we have

$$\sin x \cosh y = e^{\alpha} \cos \beta \quad ...(1)$$

and $$\cos x \sinh y = e^{\alpha} \sin \beta. \quad ...(2)$$

Squaring and adding (1) and (2), we get

$$\sin^2 x \cosh^2 y + \cos^2 x \sinh^2 y = e^{2\alpha} (\cos^{2\beta} + \sin^2 \beta)$$

or $$\frac{1}{2}(1 - \cos 2x)\cosh^2 y + \frac{1}{2}(1 + \cos 2x) \sinh^2 y = e^{2\alpha}$$

or $$\frac{1}{2}\left(\cosh^2 y + \sinh^2 y\right) - \frac{1}{2}\cos 2x\left(\cosh^2 y - \sinh^2 y\right) = e^{2\alpha}$$

or $$\frac{1}{2}\cosh 2y - \frac{1}{2}\cos 2x = e^{2\alpha}$$

or $$\frac{1}{2}\left(\cosh 2y - \cos 2x\right) = e^{2\alpha}. \quad ...(A)$$

$\therefore$ $$2\alpha = \log \frac{1}{2}\left(\cosh 2y - \cos 2x\right),$$

or $\alpha = \frac{1}{2}\log\frac{1}{2}\left(\cosh 2y - \cos 2x\right)$, which proves the result (i).

From the result (A), we have

$$\cosh 2y - \cos 2x = 2e^{2\alpha}$$

or $$\cosh 2y = \cos 2x + 2e^{2\alpha}$$

or $$2 \cosh 2y = 2 \cos 2x + 4e^{2\alpha}$$

or $2 \cos 2x = 2 \cosh 2y - 4e^{2\alpha} = e^{2y} + e^{-2y} - 4e^{2\alpha}$, which proves the result (ii).

Example 5(a):

Find the principal and general value of log (–1 + i).

Solution:

Let $-1 + i = r (\cos \theta + i \sin \theta)$,

so that $r \cos \theta = -1$ and $r \sin \theta = 1$.

Squaring and adding, we have

$r^2 = 2$, *i.e.,* $r = \sqrt{2}$.

Now $\cos \theta = -\frac{1}{\sqrt{2}}$ and $\sin q = \frac{1}{\sqrt{2}}$,

so that $\theta = \frac{3}{4}\pi$.

$$\therefore \; -1 + i = \sqrt{2}\left(\cos\frac{3}{4}\pi + i \sin\frac{3}{4}\pi\right) = \sqrt{2e^{(3\pi/4)i}}$$

$\therefore$ the general value is

$$\log(-1+i) = \log\left\{\sqrt{2e^{(3\pi/4)i}\,e^{2n\pi i}}\right\}$$

$$= \log\sqrt{2} + \frac{3}{4}\pi i + 2n\pi i = \frac{1}{2}\log 2 + \left(2n + \frac{3}{4}\right)\pi i.$$

Putting n = 0, the principal value is given by

$$\log(-1+i) = \frac{1}{2}\log 2 + \frac{3}{4}\pi i.$$

Example 5(b):

If log sin (x + iy) = a + ib, show that

(i) $\alpha = \frac{1}{2}\log\frac{1}{2}\,(\cosh 2y - \cos 2x)$,

(ii) $2\cos 2x = e^{2y} + e^{-2y} - 4e^{2\alpha}$,

(iii) $b = \tan^{-1}(\cot x \tanh y)$,

(iv) $\cos(x - \beta) = e^{2y}\cos(x + \beta)$.

Solution:

$\because$ $\log\sin(x + iy) = \alpha + i\beta$,

$\therefore$ $\sin(x + iy) = e^{\alpha + i\beta} = e^{\alpha}\,e^{i\beta}$

or $\sin x \cos iy + \cos x \sin iy = e^{\alpha}(\cos\beta + i\sin\beta)$

or $\sin x \cosh y + i\cos x \sinh y = e^{\alpha}\cos\beta + i\,e^{\alpha}\sin\beta$.

Equating real and imaginary parts, we have

$$\sin x \cosh y = e^{\alpha}\cos\beta \qquad ...(1)$$

and $$\cos x \sinh y = e^{\alpha}\sin\beta. \qquad ...(2)$$

Squaring and adding (1) and (2), we get

$$\sin^2 x \cosh^2 y + \cos^2 x \sinh^2 y = e^{2\alpha}(\cos^{2\beta} + \sin^2\beta)$$

or $$\frac{1}{2}(1 - \cos 2x)\cosh^2 y + \frac{1}{2}(1 + \cos 2x)\sinh^2 y = e^{2\alpha}$$

or $$\frac{1}{2}\left(\cosh^2 y + \sinh^2 y\right) - \frac{1}{2}\cos 2x\left(\cosh^2 y - \sinh^2 y\right) = e^{2\alpha}$$

or $$\frac{1}{2}\cosh 2y - \frac{1}{2}\cos 2x = e^{2\alpha}$$

or $$\frac{1}{2}(\cosh 2y - \cos 2x) = e^{2\alpha}. \qquad ...(3)$$

$\therefore \quad 2\alpha = \log \frac{1}{2}(\cosh 2y - \cos 2x),$

or $\quad \alpha = \frac{1}{2}\log\frac{1}{2}(\cosh 2y - \cos 2x)$, which proves the result (i).

From the result (A), we have

$$\cosh 2y - \cos 2x = 2e^{2\alpha}$$

or $\quad \cosh 2y = \cos 2x + 2e^{2\alpha}$

or $\quad 2 \cosh 2y = 2 \cos 2x + 4e^{2\alpha}$

or $\quad 2 \cos 2x = 2 \cosh 2y - 4e^{2\alpha} = e^{2y} + e^{-2y} - 4e^{2\alpha}$, which proves the result (ii).

Example 5(c):

Separate log sin (x + iy) into real and imaginary parts.

Solution:

We have

$\log \sin (x + iy) = \log (\sin x \cos iy + \cos x \sin iy)$

$= \log (\sin x \cosh y + i \cos x \sinh y)$

$$= \frac{1}{2}\log\left(\sin^2 x \cosh^2 y + \cos^2 x \sinh^2 y\right) + i \tan^{-1}\left(\frac{\cos x \sinh y}{\sin x \cosh y}\right) + 2n\pi i$$

$$= \frac{1}{2}\log\frac{1}{2}(1 - \cos 2x)\cosh^2 y + \frac{1}{2}(1 + \cos 2x)\sinh^2 y$$
$$+ i \tan^{-1} (\cot x \tanh y) + 2npi$$

$$= \frac{1}{2}\log\left\{\frac{1}{2}\left(\cosh^2 y + \sinh^2 y\right) - \frac{1}{2}\cos 2x\left(\cosh^2 y - \sinh^2 y\right)\right\}$$
$$+ i \tan^{-1} (\cot x \tanh y) + 2npi$$

$$= \frac{1}{2}\log\left(\frac{1}{2}\cosh 2y - \frac{1}{2}\cos 2x\right) + i\left[2n\pi + \tan^{-1}(\cot x \tanh y)\right]$$

$$= \frac{1}{2}\log\left[\frac{1}{2}(\cosh 2y - \cos 2x)\right] + i\left[2n\pi + \tan^{-1}(\cot x \tanh y)\right],$$

which is of the form P + iQ.

To find out the principal value, we put n = 0 in the above result. So we have

$$\log \sin (x + iy) = \frac{1}{2}\log\left[\frac{1}{2}(\cosh 2y - \cos 2x)\right] + i \tan^{-1}(\cot x \tanh y).$$

Example 6:

Prove that $i^i = e^{-(4n+1)\pi/2}$.

Solution:

We have $i^i = e^{i \log i}$, by def. of a^z

$= \exp(i \log i) = \exp[i(\log i + 2n\pi i)]$. ...(1)

Now $\log i = \log\left(\cos\frac{1}{2}\pi + i\sin\frac{1}{2}\pi\right)$ $\left[\because i = \cos\frac{1}{2}\pi + i\sin\frac{1}{2}\pi\right]$

$\therefore \log e^{i\pi/2} = i\pi/2$.

Substituting for log i in (1), we have

$$i^i = \exp\left[i\left(i\frac{1}{2}\pi + 2n\pi i\right)\right] = \exp\left[i^2\left(2n + \frac{1}{2}\right)\pi\right]$$

$= \exp[-(4n+1)\pi/2] = e^{-(4n+1)\pi/2}$ ...(2)

Note 1: Putting n = 0 in (2), we get the principal value of $i^i = e^{-\pi/2}$.

2: Putting n = 0, 1, 2, ... in (2), the various values of i^i are $e^{-\pi/2}, e^{-5\pi/2}, e^{-9\pi/2}, e^{-13\pi/2}, \ldots$

which is a geometrical progression with common ratio $e^{-2\pi}$.

Example 7:

If $i^{\alpha + i\beta} = e^x(\cos y + i \sin y)$, *then prove that*

$x = -\frac{1}{2}(4n+1)\pi\beta$ *and* $y = \frac{1}{2}(4n+1)\pi\alpha$.

Solution:

We have

$e^x(\cos y + i\sin y) = i^{(\alpha + i\beta)}$. [given]

Now $i^{(\alpha+i\beta)} = \exp[(\alpha + i\beta)\log i]$

$= \exp[(a + i\beta)(\log i + 2n\pi i)]$

$= \exp[(\alpha + i\beta)(\log e^{i\pi/2} + 2n\pi i)]$ $\left[\because e^{i\pi/2} = \cos\frac{1}{2}\pi + i\sin\frac{1}{2}\pi = i\right]$

$$= \exp\left[(\alpha + i\beta)\left(i\frac{1}{2}\pi + 2n\pi i\right)\right] = \exp[(i\alpha - \beta)(4n+1)\pi/2]$$

$= e^{(i\alpha - \beta)(4n+1)\pi/2} = e^{-(4n+1)\pi\beta/2}\, e^{i(4n+1)\pi\alpha/2}$

$= e^{-(4n+1)\pi\beta/2}[\cos\{(4n+1)\pi\alpha/2\} + i\sin\{(4n+1)\pi\alpha/2\}]$.

$\therefore$ $e^x (\cos y + i \sin y) = e^{-(4n+1)\pi\beta/2} [\cos \{(4m + 1) \pi\alpha/2\} +$
$i \sin \{(4n + 1\} \pi\alpha/2\}]$.

Equating real and imaginary parts, we get

$e^x \cos y = e^{-(4n+1) pb/2} \cos \{(4n + 1) \pi\alpha/2\}$ ...(1)

and $e^x \sin y = e^{-(4n+1)\pi\beta/2} \sin \{(4n + 1) \pi\alpha/2\}$. ...(2)

Squaring and adding (1) and (2), we get

$(e^x)^2 = [e^{-(4n+1)\pi\beta/2}]^2$

or $e^x = e^{-(4n+1)\pi\beta/2}$

or $x = -(4n + 1) \pi\beta/2$.

Dividing (2) by (1), we get

$\tan y = \tan \{(4n + 1) \pi\alpha/2\}$

or $y = (4m + 1) \pi\alpha/2$.

Example 8:

If $i^{i...ad inf.} = A + iB$, principal values only being considered, prove that

(i) $\tan \frac{1}{2} \pi A = B/A$, *and*

(ii) $A^2 + B^2 = e^{-\pi\beta}$.

Solution:

We have $i^{i^{i...ad inf.}} = A + iB$.

$\therefore$ $i^{A+iB} = A + iB$

or $\exp [(A + iB) \log i] = A + iB$, taking the principal value of i^{A+iB}

or $\exp [(A + iB) \log \left(\cos\frac{1}{2}\pi + i \sin\frac{1}{2}\pi\right) = A + iB$

or $\exp [(A + iB) \log e^{i\pi/2}] = A + iB$

or $\exp \left[(A + iB) i\frac{1}{2}\pi\right] = A + iB$

or $\exp\left[-\frac{1}{2}\pi B + i\frac{1}{2}\pi A\right] = A + iB$

or $e^{-B\pi/2} e^{\pi Ai/2} = A + iB$

or $e^{-B\pi/2}\left(\cos\frac{1}{2}\pi A + i\sin\frac{1}{2}\pi A\right) = A + iB$

Equating real and imaginary parts on both sides, we have

$$e^{-B\pi/2}\cos\frac{1}{2}\pi A = A, \qquad ...(1)$$

and $$e^{-B\pi/2}\sin\frac{1}{2}\pi A = B.$$

Dividing (2) by (1), we have

$$\tan\frac{1}{2}\pi A = \frac{B}{A}.$$

Squaring and adding (1) and (2), we get

$$\left(e^{-B\pi/2}\right)^2\left(\cos^2\frac{1}{2}\pi A + \sin^2\frac{1}{2}\pi A\right) = A^2 + B^2$$

or $$e^{-B\pi} = A^2 + B^2.$$

Example 9(a):

IF $p^{a+ib} = (x+iy)^{m+in}$ and the principal values are considered, prove that

(i) $a = \frac{1}{2}m\log_p(x^2+y^2) - n\tan^{1}\left(\frac{y}{x}\right)\log_p e,$

(ii) $\log_p(x^2+y^2) = \frac{2(\alpha m+\beta n)}{m^2+n^2}.$

Solution:

We have $p^{\alpha+i\beta} = (x+iy)^{m+in}$.

Taking logarithm of both sides, we have

$(\alpha+i\beta)\log_e p = (m+in)\log_e(x+iy)$

$$= (m+n)\left[\frac{1}{2}\log_e\left(x^2+y^2\right) + i\tan^{-1}\left(\frac{y}{x}\right)\right],$$

$$= \left[\frac{1}{2}m\log_e\left(x^2+y^2\right) - n\tan^{-1}\left(\frac{y}{x}\right)\right]$$

$$+ i\left[\frac{1}{2}n\log_e\left(x^2+y^2\right) + m\tan^{-1}\left(\frac{y}{x}\right)\right]$$

Equating real and imaginary parts on both sides, we get

$$\left[\frac{1}{2}n\log_e\left(x^2+y^2\right)+m\tan^{-1}\left(\frac{y}{x}\right)\right] \quad ...(1)$$

and $\beta \log_e p = \frac{1}{2}m\log_e\left(x^2+y^2\right)+m\tan^{-1}\left(\frac{y}{x}\right).$ ...(2)

(i) From (1), we have

$$\alpha \log_e p = \frac{1}{2}m \log_e (x^2 + y^2) - n \tan^{-1}\left(\frac{y}{x}\right)$$

or $(\alpha \log_e p) \, . \, (\log_p e) = \left[\frac{1}{2}m\log_e\left(x^2+y^2\right)-n\tan^{-1}\left(\frac{y}{x}\right)\right].\log_e e$

multiplying both sides by $\log_p e$

or $\alpha = \frac{1}{2}m\log_p\left(x^2+y^2\right)-n\tan^{-1}\left(\frac{y}{x}\right).\log_p e.$

[$\because \log_b \alpha \times \log_a \beta = 1; \log_a m \times \log_b \alpha = \log_b m$]

$\therefore \; \alpha = \frac{1}{2}m\log_p\left(x^2+y^2\right)-n\tan^{-1}\left(\frac{y}{x}\right)\log_p e.$ **Proved.**

(ii) Multiplying (1) by m and (2) by n and adding, we get

$$(\alpha m + \beta n)\log_e p = \frac{1}{2}(m^2 + n^2)\log_e (x^2 + y^2)$$

or $(\alpha m + \beta n)\log_e p \times \log_e = \frac{1}{2}(m^2 + n^2)\log_e (x^2 + y^2) \times \log_p e,$

multiplying both sides by $\log_p e$

or $(\alpha m + \beta n) \, . \, 1 = 1/2\,(m^2 + n^2)\log_p (x^2 + y^2)$

or $\log_p (x^2 + y^2) = \frac{2\left(\alpha m+\beta n\right)}{\left(m^2+n^2\right)}.$

Example 9(b):

If $(a + ib)^p = m^{x + iy}$, then prove that

$$\frac{y}{x} = \frac{2\tan^{-1}\left(\frac{b}{a}\right)}{\log\left(a^2+b^2\right)},$$

where only principal values are considered.

Solution:

Given that $(a + ib)^p = m^{(x + iy)}$.

Taking logarithm of both sides, we have

$$p \log (a + ib) = (x + iy) \log m$$

or $$p\left[\frac{1}{2}\log\left(a^2+b^2\right)+i\tan^{-1}\left(\frac{b}{a}\right)\right] = x \log m + iy \log m.$$

Equating real and imaginary parts, we get

$$\frac{1}{2}p\log\left(a^2+b^2\right) = x \log m \qquad ...(1)$$

and $$p\tan^{-1}\left(\frac{b}{a}\right) = y \log m \qquad ...(2)$$

Dividing (2) by (1), we get

$$\frac{y}{x} = \frac{p\tan^{-1}\left(\frac{b}{a}\right)}{\left(\frac{p}{2}\right)\log\left(a^2+b^2\right)} = \frac{2\tan^{-1}\left(\frac{b}{a}\right)}{\log\left(a^2+b^2\right)}.$$

Example 10:

Express log $(1 + i)^{(1 - i)}$ in the form A + iB.

Solution:

We have

$\log (1 + i)^{(1 - i)} = (1 - i) \log (1 + i)$

$= (1-i)\left[\frac{1}{2}\log\left(1^2+1^2\right)+i\tan^{-1}1\right] = (1-i)\left[\frac{1}{2}\log 2+i\frac{1}{4}\pi\right]$

$= \left(\frac{1}{2}\log 2+\frac{1}{4}\pi\right)+i\left(\frac{1}{4}\pi+\frac{1}{2}\log 2\right)$, which is of the form A + iB.

Example 11:

Find the general value of log (–i).

Solution:

We have $-i = \left(\cos\frac{1}{2}\pi - i\sin\frac{1}{2}\pi\right) = e^{-i\pi/2}$

so that $\log (-i) = \log e^{-i\pi/2} = -i\pi/2$, giving the principal value.

$\therefore \log (-i) = \log (-i) + 2n\pi i = -\left(\frac{i\pi}{2}\right)+2n\pi i = \frac{1}{2}(4n-1)\pi i.$

Example 12:

Find the general value of log (–3).

Solution:

Let $-3 = -3 + i0 = r(\cos\theta + i\sin\theta)$,

so that $-3 = r\cos\theta$ and $0 = r\sin\theta$.

These give $r^2 = 9$ *i.e.*, $r = 3$. Putting $r = 3$, we get

$\cos\theta = -1$ and $\sin\theta = 0$, giving $\theta = \pi$.

$\therefore\ -3 = 3(\cos\pi + i\sin\pi) = 3.e^{i\pi}$.

$\therefore\ \log(-3) = \log\{3.e^{i\pi}.e^{2n\pi i}\}$ $\qquad (\because e^{2n\pi i} = 1)$

$= \log 3 + \log e^{(2n\pi + \pi)i} = \log 3 + (2n+1)\pi i.$

The principal value of log (–3) *i.e.*, log (–3) is obtained by putting n = 0 in the above result. Thus $\log(-3) = \log 3 + i\pi$.

Example 13:

Prove that

$$\log\left(\frac{1}{1-e^{i\alpha}}\right) = \log\left(\frac{1}{2}\operatorname{cosec}\frac{\alpha}{2}\right) + i\left(\frac{\pi}{2} - \frac{\alpha}{2}\right).$$

Solution:

We have

$$\log\frac{1}{1-e^{i\alpha}} = \log\frac{1}{1-\cos\alpha - i\sin\alpha} \qquad [\because e^{i\alpha} = \cos\alpha + i\sin\alpha]$$

$$= \log\frac{1}{2\sin^2\frac{1}{2}\alpha - 2i\sin\frac{1}{2}\alpha\cos\frac{1}{2}\alpha} = \log\frac{1}{2\sin\frac{1}{2}\alpha\left(\sin\frac{1}{2}\alpha - i\cos\frac{1}{2}\alpha\right)}$$

$$= \log\frac{1}{2\sin\frac{1}{2}\alpha\left\{\cos\left(\frac{1}{2}\pi - \frac{1}{2}\alpha\right) - i\sin\left(\frac{1}{2}\pi - \frac{1}{2}\alpha\right)\right\}}$$

$$= \log\frac{1}{2\sin\frac{1}{2}\alpha\ e^{-i(\pi/2-\alpha/2)}} \qquad [\because e^{-i\theta} = \cos\theta - i\sin\theta]$$

$$= \log\left[\left(\frac{1}{2}\operatorname{cosec}\frac{1}{2}\alpha\right).e^{i(\pi/2-\alpha/2)}\right]$$

$$= \log\left(\frac{1}{2}\operatorname{cosec}\frac{1}{2}\alpha\right) + \log e^{i(\pi/2-\alpha/2)}$$

$$= \log\left(\frac{1}{2}\operatorname{cosec}\frac{1}{2}\alpha\right) + i\left(\frac{1}{2}\pi - \frac{1}{2}\alpha\right).$$

EXERCISES

1. Prove that :

$$\log\left(\frac{a+ib-x}{a+ib+x}\right)=\frac{1}{2}\log\left\{\frac{(a-x)^2+b^2}{(a+x)^2+b^2}\right\}+i\tan^{-1}\left[\frac{b}{a-x}-\tan^{-1}\frac{b}{a+x}\right].$$

2. If $\frac{(1+i)^{p+iq}}{(1-i)^{p-iq}}=\alpha+i\beta$, prove that one value of $\tan^{-1}\left(\frac{\beta}{\alpha}\right)$ is $\left(\frac{\pi p}{2}\right)+q\log 2$.

3. If $[\cos(\theta - i\phi)]^{x+iy} = A + iB$ and principal values are taken into consideration, then

$$\tan^{-1}\left(\frac{B}{A}\right)=\frac{1}{2}\, y\log(\cosh^2\phi-\sin^2\theta)+x\tan^{-1}(\tan\theta\tanh\phi).$$

4. If $\left\{\frac{a+x+iy}{a-x-iy}\right\}^{\lambda+\mu i} = X + iY$, prove that one of the values of $\tan^{-1}\left(\frac{Y}{X}\right)$ is $\lambda\tan^{-1}\left\{\frac{2ay}{a^2-x^2-y^2}\right\}+\frac{\mu}{2}\log\left\{\frac{(a+x)^2+y^2}{(a-x)^2+y^2}\right\}$.

5. Prove that $\log(1+i)=\frac{1}{2}\log 2+i\left(2n\pi+\frac{1}{4}\pi\right)$.

6. Prove that $\log(-5) = \log 5 + (2n\pi + \pi)\, i$.

7. Find the general value of log i.

8. Find the general value of log (–i).

9. Find the general value of $\log\sqrt{i}$.

10. Show that $\log(1+e^{i\theta})=\log\left(2\cos\frac{1}{2}\theta\right)+\frac{1}{2}i\theta$, if $-\pi<\theta<\pi$.

11. Prove that $\log\left(\frac{a+ib}{a-ib}\right)=2i\tan^{-1}\left(\frac{b}{a}\right)$.

12. Show that $i\log\frac{x-i}{x+i}=\pi-2\tan^{-1}x$.

13. Show that $\tan\left(i\log\frac{a-ib}{a+ib}\right)=\frac{2ab}{a^2-b^2}$.

14. Prove that $\log(1 + i\tan\alpha) = \log\sec\alpha + \alpha i$.

15. Prove that sin (log i^i) = – 1.

16. If u = log tan $\left(\frac{\pi}{4}+\frac{\theta}{2}\right)$, prove that

 (i) $\tanh\frac{u}{2} = \tan\frac{\theta}{2}$.

 (ii) $\theta = -i\log\tan\left(\frac{\pi}{4}-\frac{iu}{2}\right)$.

17. Prove that log (1 + cos 2θ + i sin 2θ) = log (2 cos θ) + iθ, iϕ – π < θ ≤ π.

18. Prove that log (iβ) = log |β| ± i 1/2π, + or – sign being taken as β is +ive or –ive.

19. Separate into real and imaginary parts: log sin (x + iy).

20. Prove that

$$\log\left[\frac{\sin(x+iy)}{\sin(x-iy)}\right] = 2i\tan^{-1}(\cot x\tanh y).$$

21. Prove that log cos (x + iy) = 1/2 log 1/2(cosh 2y + cos 2x) – i $\tan^{-1}$ (tan x tanh y).

22. Prove that $\log\left\{\frac{\cos(x-iy)}{\cos(x+iy)}\right\} = 2i\tan^{-1}(\tan x\tanh y)$.

23. Show that one of the values of $\log\frac{(1+i)\left(1+i\sqrt{3}\right)}{\sqrt{3}+i}$ is $\frac{1}{2}\log 2 + i\frac{5}{12}\pi$.

24. If tan log (x + iy) = a + ib, where $a^2 + b^2 \neq 1$,

 prove that $\tan\{\log(x^2 + y^2)\} = \frac{2a}{1-a^2-b^2}$.

25. If log log (x + iy) = p + iq, prove that

$$y = x\tan[\tan q\log(x^2 + y^2)^{1/2}].$$

26. If log log log (α + iβ) = p + iq, prove that

 (i) $e^{e^p\cos q}.\sin\left(e^p\sin q\right) = \frac{1}{2}\text{loig}\left(\alpha^2+\beta^2\right)$,

 (ii) $e^{e^p\cos q}.\sin\left(e^p\sin q\right) = \tan^{-1}\left(\frac{\beta}{\alpha}\right)$.

27. Prove that $\log\frac{(a-b)+i(a+b)}{(a+b)+i(a-b)} = i\left\{2n\pi+\tan^{-1}\frac{2ab}{a^2-b^2}\right\}$.

28. If $(a_1 + ib_1)(a_2 + ib_2)\ldots(a_n + ib_n) = A + iB$, prove that

$$\tan^{-1}\left(\frac{b_1}{a_1}\right)+\tan^{-1}\left(\frac{b_2}{a_2}\right)+\ldots+\tan^{-1}\left(\frac{b_n}{a_n}\right)=\tan^{-1}\frac{B}{A},$$

and $\left(a_1^2+b_1^2\right)\left(a_2^2+b_2^2\right)\ldots\left(a_n^2+b_n^2\right)=A^2+B^2.$

29. If (1 + i) (1 + 2i) (1 + 3i) ... (1 + ni) = A + iB, show that 2.5.10...$(1 + n^2) = (A^2 + B^2)$.

30. If $i^{\alpha + i\beta} = \alpha + i\beta$, show that $\alpha^2 + \beta^2 = e^{-(4n+1)\pi\beta}$.

31. Prove that $i^a = \cos\left[\left(2m+\frac{1}{2}\right)\pi a\right]+i\sin\left[\left(2m+\frac{1}{2}\right)\pi a\right]$.

32. If $i^{i^{i}} = \cos\theta + i\sin\theta$, prove that $\theta=\left(2m+\frac{1}{2}\right)\pi\exp\left[-\left(2n+\frac{1}{2}\right)\pi\right]$.

33. If $(i^i)^i = \cos\theta - i\sin\theta$, show that $\theta=\frac{1}{2}\pi(4n-1)$.

34. Show that the sum of the moduli of the values of $(1 + i)^{1+i}$, which are less than unity is

$$\left(\frac{1}{\sqrt{2}}\right).e^{3\pi/4}.\text{cosech }\pi.$$

35. Show that the principal value of $(a + ib)^{p+iq}/(a - ib)^{p-iq}$ is $\cos 2(p\alpha + q\log r) + i\sin 2(p\alpha + q\log r)$, where $r=\sqrt{\left(a^2+b^2\right)}$ and $\alpha=\tan^{-1}\left(\frac{b}{a}\right)$.

36. Prove that the real part of the principal value of $(i)^{\log(1+i)}$ is $\exp\left(-\frac{\pi^2}{8}\right)\times\cos\left(\frac{1}{4}\pi\log 2\right)$.

37. Prove that $(x + ix\tan y)^{\log(x\sec y) - iy}$ is real, when only principal values are considered.

38. Prove that the principal value of $(a + ib)^{c+id}$ is wholly real or wholly imaginary according as

$$\frac{1}{2}d\log\left(a^2+b^2\right)+c\tan^{-1}\left(\frac{b}{a}\right)$$

is an even or an odd multiple of $\frac{1}{2}\pi$.

In case it is wholly real, prove that $(a + ib)^{c+id} = (a^2 + b^2)^{(c^2+d^2)/2c}$.

39. Prove that the general value of $(1 + i\tan\alpha)^{-i}$ is $\exp(\alpha + 2m\pi).[\cos(\log\cos\alpha) + i\sin(\log\cos\alpha)]$.

4

Parabola

INTERSECTION OF A STRAIGHT LINE AND A PARABOLA

Let the equation of the straight line be given as

$$y = mx + c \qquad ...(i)$$

and the eqn. of parabola is

$$y^2 = 4ax. \qquad ...(ii)$$

The point of intersection of the curves (i) and (ii) can be obtained as

$$(mx + c)^2 = 4ax$$

$$\Rightarrow \qquad m^2x^2 + 2x(cm - 2a) + c^2 = 0. \qquad ...(iii)$$

since the equation (iii) is quadratic in x, it gives two points of intersection between a straight line and a parabola. These points can be real and different, real and coincident or imaginary according as the roots of the equation (iii) are real and different, real and equal or imaginary.

PROPERTIES OF THE PARABOLA

The tangent at any point P on the parabola bisects the angle between the focal chord through P and the perpendicular from P on the directrix.

Proof:

Let P $(at^2, 2at)$ be any point on the parabola, PT is the tangent to the parabola $y^2 = 4ax$, s is the focus, HZ' is the directrix and PM is perpendicular from P on ZZ'.

Equation of tangent PT is given by $ty = x + at^2$ given by

$$\therefore \text{ Slope of PT} = \frac{1}{t}$$

Also slopes of PM and SP and respectively 0 and

$$\frac{2at-0}{at^2-a} \text{ or } 0 \text{ and } \frac{2t}{t^2-1}$$

If θ and oϕ be the angles MPT and SPT respectively, then

We have $\tan\theta = \dfrac{\frac{1}{t}-0}{1+\left(\frac{1}{t}\right),0} = \dfrac{1}{t}$

and $$\tan\phi = \frac{\frac{2t}{t^2-1}-\frac{1}{t}}{1+\frac{2t}{(t^2-1)}.\frac{1}{t}} = \frac{1}{t}.$$

∴ tan θ = tan ϕ or θ = ϕ

which is the result

Show that locus of the foot of perpendicular from the focus on any tangent to the parabola is the tangent at the vertex.

Proof:

Equation of any tangent to the parabola $y^2 = 4ax$ is

$$y = mx + a/m. \quad \text{...(i)}$$

The equation of any line through the focus (a, 0) and perpendicular to equation is (i) is given as

$$y = -\frac{1}{m}x + a/m. \quad \text{...(ii)}$$

Subtracting (i) from (ii), we have

$$\left(m+\frac{1}{m}\right)x = 0$$

⇒ $x = 0$,

which is the tangent at the vertex. Hence the result

The tangents at the extremities of a focal chord intersect at right angles on the directrix.

Proof:

Let $P(at_1^2, 2at_1)$ and $Q(at_2^2, 2at_2)$ be the extremities of the focal chord PSQ. Equation of PQ is given by

$$y - 2at_1 = \frac{2at_1 - 2at_2}{at_1^2 - at_2^2}(x - at_1^2).$$

$$y - 2at_1 = \frac{2}{t_1 + t_2}(x - at_1^2).$$

Since the chord passes through the focus (a, 0) then we have

$$0 - 2at_1 = \frac{2}{t_1+t_2}(a - at_1^2)$$

$$-2at_1(t_1 + t_2) = 2a - 2at_1^2$$

or we have $t_1t_2 = -1.$ ...(i)

Equation of tangents at P and Q are

Given by $t_1y = x + at_1^2$ ...(ii)

and $t_2y = x + at_2^2$ ...(iii)

The point of intersection of these tangents is obtained by solving equation (ii) and (iii) simultaneously. This gives

$$x = at_1t_2.$$ ...(iv)

The slope of (ii) is $\frac{1}{t_1}$ and of (iii) is $\frac{1}{t_2}$. The product of these two slopes is $\frac{1}{t_1t_2}$. ...(v)

Using the result $t_1t_2 = -1$ (iv) gives $x = -a$, *i.e.*, the tangent meet at the directrix. Thus (v) gives the product of the slopes of these tangents as -1, *i.e.*, the tangents are at right angles.

The portion of a tangent to a parabola cut off between the directrix and the curve subtends a right angle at the focus.

Proof:

Let P $(at^2, 2at)$ be any point on the parabola. The equation of tangent at P is given by

$$ty = x + at^2.$$ ...(i)

The co-ordinates of T are obtained by solving equation (i) and $x = -a$, the equation of the directrix. Which gives

$$ty = -a + at^2$$

$$\Rightarrow \quad y = \frac{(t^2-1)}{t}a.$$

∴ The co-ordinates of T are as follows:

$$\left(-a, \frac{a(t^2-1)}{t}\right).$$

Slope of PS $= m_1 = \frac{2at}{at^2 - a} = \frac{2at}{a(t^2-1)}.$ ...(iii)

Slope of TS $m_2 = \dfrac{a(t^2-1)}{\frac{t}{-2a}}$. ...(iv)

Now $m_1 \times m_2 = -1$ hence the result

If SY be perpendicular to the tangent at a point P of a parabola, prove that $SY^2 = AS.SP$.

Proof:

Take any point P $(at^2, 2at)$.

Equation of tangent at P is given by

$$ty = x + at^2. \quad ...(i)$$

Equation of SY passing through S (a, 0) and perpendicular to PT is given by

$$y = (-t)(x-a)$$

i.e. $$tx + y = at. \quad ...(ii)$$

Solving (i) and (ii), we obtain

$$x = 0 \text{ and } y = at.$$

$\therefore$ the co-ordinates of Y are (0, at).

$$\therefore \quad SY^2 = a^2 + a^2 t^2$$
$$= a^2 (1 + t^2)$$
$$= a.(a + at^2) = \angle S.SP.$$

Hence $SY^2 = AS.\ SP$ hence the result

The orthocentre of any triangle formed by three tangent to a parabola lies on the directrix.

Proof:

Let the point be $P(at_1^2, 2at_1)$, $Q(at_2^2, 2at_2)$ and $R(at_3^2, 2at_3)$. The equation of the tangents at P, Q and R are

$$t_1y = x + at_1^2 \quad ...(i)$$
$$t_2y = x + at_2^2 \quad ...(ii)$$

and $$t_3y = x + at_3^2. \quad ...(iii)$$

The lines (i) and (ii) intersect at $\{at_1t_2, a(t_1 + t_2)\}$.

The equation of the perpendicular from this point on (iii) is given by

$$t_3x + y = a(t_1 + t_2 + t_1t_2t_3). \quad ...(iv)$$

Similarly, the equation of the perpendicular on (i) from the point of intersection of (ii) and (iii) is given by

$$t_1x + y = a(t_2 + t_3 + t_1t_2t_3). \qquad ...(v)$$

The x-co-ordinate of the orthocentre which is the point of intersection of (iv) and (v) is–a. The orthocentre, therefore, lies on the directrix. Hence the result

LOCUS OF THE MIDDLE POINTS OF A SYSTEM OF PARALLEL CHORDS OF A PARABOLA

Let the equation of the parabola be

$$y^2 = 4ax \qquad ...(i)$$

Let the slope of the system of parallel chords be m. Let us take any chord whose middle point is (h, k). Then the equation of the chord is

$$y - k = \frac{2a}{k}(x - h).$$

But the slope of the chord is m, therefore, we have

$$m = \frac{2a}{k}$$

or $$k = \frac{2a}{m}$$

Hence the locus is $y = \frac{2a}{m}$, which is a straight line parallel to the axis of the parabola. This locus is called diameter.

TRACING OF THE PARABOLA

The parabola can be traced according to the following steps

(i) Changing of y to –y does not change the equation of the parabola. This implies that the curve is symmetrical about x-axis.

(ii) Since the negative value of x makes the value of y (obtained from the equation of the parabola) imaginary, no part of the curve can lie to the left of the y-axis

(iii) The curve passes through origin.

(iv) The value of y increases as x increases. Using the above steps, we plot the curve and shape of parabola is shown in the figure

POSITION OF A POINT RELATIVE TO A PARABOLA

The point (x_1, y_2) lies outside, on or inside the parabola $y^2 = 4ax$ according as the expression $y_1^2 - 4ax_1$ is positive, zero or negative.

If (x_1, y_1) is a given point and $y_1^2 - 4ax_1 = 0$, then by definition (x_1, y_2) lies on the parabola. In case $y_1^2 - 4ax_1$ is non zero, we proceed as follows :

Let $P(x_1, y_1)$ be a point lying outside the parabola and PM be perpendicular to x-axis. This gives

$$PM > QM$$

$$\Rightarrow PM^2 - QM^2 > 0$$

Now $PM^2 = y_1$, and $QM^2 = 4ax_1$, as the co-ordinate of Q is (x_1, QM) and Q lies on the parabola.

Substituting the values of PM and QM, we get

$$y_1^2 - 4ax_1 > 0$$

which is the required condition for the point $P(x_1, y_1)$ to be outside the parabola. Similarly, the condition for $P(x_1, y_1)$ to lie inside the parabola can be obtained.

TANGENTS AT THE EXTREMITY OF A DIAMETER

The equation of a diameter of the parabola bisects system of parallel chords having slope m is given by

$$y = \frac{2a}{m}.$$

The intersection of this diameter and the parabola $y^2 = 4ax$ is

$$\frac{4a^2}{m^2} = 4ax$$

$$\Rightarrow \quad x = \frac{a}{m^2}$$

Therefore the co-ordinates of the extremity is $(a/m^2. 2a/m)$.

Equation of the tangent at the point is given by

$$y - \frac{2a}{m} = 2a(x + a/m^2)$$

$$\Rightarrow \quad y = mx + a/m,$$

which is parallel to the given system of chords. This proves that the tangents at the extremity of a diameter of parabola is parallel to the chords which that diameter bisects.

TO SHOW THAT TANGENTS AT THE ENDS OF A CHORD OF A PARABOLA MEET ON THE DIAMETER WHICH BISECTS THE CHORD

Let $P(at_1^2, 2at_1)$ and $Q(at_2^2, 2at_2)$ be the ends of a chord of the given parabola $y^2 = 4ax$. Then

The slope of PQ is $\frac{2}{t_1 + t_2}$.

$\therefore$ The equation of the diameter bisecting the chord is given by

$$y = a(t_1 + t_2).$$

The points of intersection of the tangents at P and Q is $[a\, t_1t_2, a(t_1 + t_2)]$, which is obtained by solving the equation

$$t_1y = x + at_1^2$$

$$t_2y = x + at_2^2.$$

Since y-co-ordinate is $a(t_1 = t_2)$, the result follows.

PARAMETRIC CO-ORDINATES

The point given by

$$x = at^2,$$

$$y = 2at \qquad ...(i)$$

lies on the parabola for all values of t. The equations (i) are known as the parametric equations to the parabola and the point $(at^2, 2at)$ on the parabola is known as the point t.

TANGENT AND NORMAL AT THE POINT $(at^2, 2at)$

We know that the equation of tangent and normal at (x_1, y_1) to the parabola is given by the following equation

Tangent $yy_1 = 2a(x + x_1)$

Normal $y - y_1 = \dfrac{-y_1}{2a}(x - x_1).$

Substituting $x_1 = at^2$, $y_1 = 2at$, the corresponding equations of tangent and normal becomes

$$2aty = 2a(x + at^2)$$

$$y - 2at = -t(x - at^2)$$

$$\Rightarrow \quad ty = x + at^2 \qquad ...(i)$$

$$y + tx = 2at + at^3 \qquad ...(ii)$$

Equations (i) and (ii) are the required equations of tangent and normal to the parabola at the point $(at^2, 3at)$

CONDITION THE LINE $y = mx + c$ IS TANGENT TO THE PARABOLA $y^2 = 4ax$

The points of intersection of the line

$$y = mx + c \qquad ...(i)$$

and parabola $y^2 = 4ax$ ...(ii)

are given by the equation

$$(mx + c)^2 = 4ax$$

$$\Rightarrow \quad m^2x^2 + 3x\,(mc - 2a) + c^2 = 0. \qquad \text{...(iii)}$$

The equation (iii) being a quadratic in x gives two values of x. In case the line y = mx + c is to be tangent to the parabola, the equation (iii) should give only one point of intersection, *i.e.*, the discriminant of equation must be zero. Thus we have

$$4\,(mc - 2a)^2 - 4m^2c^2 = 0$$

$$\Rightarrow \quad 4m^2c^2 - 4amc + 4a^2 - 4m^2c^2 = 0$$

$$\Rightarrow \quad mc = c \quad \text{or} \quad c = a/m.$$

Which is the required condition. Thus y = mx + a/m is tangent to the parabola for all values of m. This is the equation of the tangent to the parabola in the slope form. Substituting c = a/m in (iii) we obtain

$$m^2c^2 + 2\,(a - 2a)\,x + a^2/m^2 = 0$$

$$(mx - a/m^2 = 0.$$

Which gives $x = a/m^2$. Equation (iii) gives y = 2a/m. Thus y = mx + a/ m is tangent to the parabola $y^2 = 4ax$ at the point $(a/m^2, 2a/m)$.

EQUATION OF TANGENT TO THE PARABOLA AT $(x_{1,}, y_1)$

Équation of the parabola is

$$y^2 = 4ax. \qquad \text{...(i)}$$

Let as take any two point $P(x_1, y_1)$ and $Q(x_2, y_2)$ on parabola equation of chord PQ is given by

$$y - y_1 = \frac{y_1 - y_2}{x_1 - x_2}(x - x_1) \qquad \text{...(ii)}$$

Since $P(x_1, y_1)$ and $Q(x_2, y_2)$ lie on parabola

$$\therefore \quad y_1^2 = 4ax_1 \qquad \text{...(iii)}$$

$$\text{and} \quad y_2^2 = 4ax_2 \qquad \text{...(iv)}$$

Subtracting (iv) from (iii) and simplifying, we get

$$(y_1 - y_2)\,(y_1 + y_2) = 4a\,(x_1 - y_2)$$

$$\text{or} \quad \frac{y_1 - y_2}{x_1 - x_2} = \frac{4a}{y_1 + y_2}.$$

Subtracting the values the values of $\frac{y_1 - y_2}{x_1 - x_2}$ in (ii), we obtain

$$\Rightarrow \qquad y - y_1 = \frac{4a}{y_1 + y_2}(x - x_1).$$

The chord PQ will be tangent to the parabola if the point Q coincides with P *i.e.* if $x_2 = x_1$ and $y_2 = y_1$. this gives

$$y - y_1 = \frac{2a}{y_1}(x - x_1)$$

$$\Rightarrow \qquad yy_1 - y_1^2 = 2ax - 2ax_1$$

$$\Rightarrow \qquad yy_1 = y_1^2 + 2ax - 2ax_1 - 2ax_1 + 2ax_1$$

$$\Rightarrow \qquad yy_1 = (y_1^2 - 4ax_1) + 2a(x + x_1)$$

$$\Rightarrow \qquad yy_1 = 2a(x + x_1). \qquad [\because (y_1^2 - 4ax_1) = 0]$$

which is the required equation of the tangent to the parabola $y^2 = 4ax$ at the point (x_1, y_1).

EQUATION OF NORMAL

Let $y^2 = 4ax$ be the equation of parabola and (x_1, y_1) be any point on it. We have to find the equation of normal at the point (x_1, y_1) is given by

Equation of a line passing through (x_1, y_1) is

$$y - y_1 = m(x - x_1) \qquad \text{...(i)}$$

Equation of a tangent to the parabola at (x_1, y_1) is given by

$$yy_1 = 2a(x + x_1) \qquad \text{...(ii)}$$

The slope of tangent is $\frac{2a}{y_1}$. Since normal is a line perpendicular to tangent, therefore the slope of normal is $\frac{y_1}{2a}$ then. Equation (i) becomes

$$y - y_1 = \frac{-y_1}{2a}(x - x_1), \qquad \text{...(iii)}$$

This is the required equation of normal at the point (x_1, y_1) to the parabola $y^2 = 4ax$.

Let $\frac{-y_1}{2a} = m$, so that $y_1 = -3am$.] Substituting y_1 in the equation $y_1^2 = 4ax_1$, we get $x_1 = am^2$. Equation (iii) can now be re-written as

$$y + 2am = m(x - am^2)$$

$$\Rightarrow \qquad y = mx - 2am - am^3. \qquad \text{...(iv)}$$

The equation (iv) is the equation of the normal to the parabola in the slope form.

CO-NORMAL POINTS

Show that three normals can be drawn from a given point to a parabola and that the algebraic sum of the ordinates of their feet is zero. We know that

The equation of normal at the point $(at^2, 2at)$ to the parabola $y^2 = 4ax$ is

$$y + tx = 2at + at^3$$

Let it passes through (h, k). Which gives

$$at^3 + 2at - th - k = 0 \quad \text{...(i)}$$

This equation, being cubic in t, will give three values of t and to each value of t there corresponds a normal. Hence from a given point three normals can be drawn to the parabola. If t_1, t_2, t_3 are the roots of (i), then ordinates of the feet of these normals are $2at_1$, $2at_2$, $2at_3$. These points are called *Co-normal Points: The sum of the ordinates of these normals is*

$$2at_1 + 2at_2 + 2at_3 = 2a(t_1 + t_2 + t_3)$$

$$= 0 \quad [\text{as } \Sigma t = 0 \text{ from equation (i)}]$$

which is the require result

Example:

Show that the distance between a tangent to the parabola $y^2 = 4ax$ and a parallel normal is a cosec θ sec^2 θ, where θ is the angle which either makes with the axis.

Solution:

The general equation of a tangent and normal are given as

$$y = mx + \frac{a}{m} \quad \text{...(i)}$$

$$y = mx - 2am - am^3. \quad \text{...(ii)}$$

Now tangent and normal meets the x-axis (y = 0) at the point $\left(\frac{-a}{m^2}, 0\right)$ and $(2a + am^2, 0)$ respectively

∴ The portion they intercept on the x-axis is given as

$$= \left[(2a + am)\left(\frac{a}{m^2}\right)\right]$$

$$= 2a + a\tan^2\theta + a\cot^2\theta \quad (\because m = \tan\theta)$$

$= a(\text{cosec}^2 \theta + \sec^2 \theta) = a \text{ cosec}^2 \theta \sec^2 \theta.$

Hence the required distance

$= a \text{ cosec}^2 \theta + \sec^2 \theta \sin \theta.$

$= a \text{ cosec}^2 \theta + \sec^2 \theta.$

FOCAL CHORD

Definition: *Any chord of the parabola which passes through focus of a parabola is called a focal chord of the parabola.*

LATUS RECTUM

A chord of the parabola passing through the focus and perpendicular to the axis is called *Latus Rectum.*

In the figure BSB' is the latus rectum.

Now $\quad BSB' = 2BS.$

The co-ordinates of B is (a, BS). Since B lies on the parabola, we must have

$$BS^2 = 4a.a = 4a^2$$

$$\Rightarrow \quad BS = 2a.$$

$$\therefore \quad BSB' = 4a,$$

which is the length of latus rectum.

PROPERTIES OF SUBTANGENT AND SUB-NORMALS

(a) The subtangent of any point on the parabola is bisected at the vertex.

Proof:

Take any point $P(at^2, 2at)$. Equation of PT is

$$ty = x + at^2.$$

It meets the axis of X when $y = 0$. This gives the co-ordinate of T as $(-at^2, 0)$. Also co-ordinate of M are $(at^2, 0)$. This means

$$AM = at^3.$$

and $\quad AT = at^2.$

$$\therefore \quad AM = AT.$$

Hence A bisects TM.

(b) The subnormal at any point of a parabola is constant and equal to the semi-latus rectum.

Proof:

Equation of normal at P is given by

$tx + y\ 2at + at^3$.

It meets the axis of X at $2at + at^2$. Therefore, the co-ordinates of G are $(2a + at^2, 0)$.

Now $\quad MG = AG - AM = 2a + at^2 - at^2 = 2a$.

Thus the subnormal is of constant length which is equal to the semi-rectum of the parabola.

SUBTANGENT AND SUBNORMAL

Let the tangent and normal at a point P of the parabola meet the axis in T and G respectively. From P draw PM perpendicular to X-axis. TM is called subtangent and MG is called subnormal of the point P.

EQUATION OF PARABOLA

Let S be the focus and ZM be the directrix. From S draw a line SAX' ⊥ to the directrix. Take A as the middle point of SB. From A draw a line AY ⊥ to BS. AS and AY are taken as the x-axis and y-axis respectively. Let AS = a. This gives AB = a and A lies on the locus. Take any point P(h, k) on the locus. Let PC be perpendicular to ZM.

Then we have SP = PC. ...(i)

∴ (i) becomes

$$\sqrt{(h-a)^2 + k^2} = (h+a).$$

Squaring, both side we get

$$h^2 - 2ah + a^2 + k^2 = h^2 + 2ah + a^2$$

or $\quad k^2 = 4ah$.

Hence the locus is given as

$$y^2 = 4ax,$$

which is required equation of the parabola with origin as the vertex, x -axis as the axis of the parabola and y-axis as the tangent at the vertex.

TANGENT THROUGH A POINT

Let $y^2 = 4ax$ be the equation of the parabola and (h, k) be the co-ordinate of the given point

We know that the line

$$y = mx + a/m \qquad \text{...(i)}$$

Is always tangent to the parabola $y^2 = 4ax$ for all values of m. If this tangent passes through (h, k), then we have

$$k = mh + a/m$$

or $$m^2h - mk + a = 0. \qquad \text{...(ii)}$$

Equation (ii) being quadratic in m gives two values of m, say m_1 and m_2. Now m_1 and m_2 lies on (ii), therefore,

We have $k = m_1h + a/m_1$

and $k = m_2h + a/m_2$.

This implies that there exists two tangents

$$y = m_1x + a/m_1$$

and $$y = m_2x + a/m_2$$

To the parabola $y^2 = 4ax$ passing through (h, k). This proves that through any point two tangents can be drawn to the parabola. The tangents can be real, coincident or imaginary depending upon the roots of the equation (ii) are real, equal or imaginary respectively. The roots of (ii) are real, coincident or imaginary according as

$$k^2 - 4ah = > = \text{ or } < 0$$

i.e., according as the point (h, k) lies outside, on or inside the parabola.

EQUATION OF A PAIR OR TANGENTS FROM AN EXTERNAL POINT

Let $P(x_1, y_1)$ be a point and let T(x, y) be any point on the line through P cutting the parabola in two points Q and R. If Q divides the line PT in some ratio λ : 1, then the co-ordinates of Q are given by

$$\left(\frac{\lambda x + x_1}{\lambda + 1}, \frac{\lambda y + y_1}{\lambda + 1}\right).$$

The point Q less on the parabola., therefore we have

$$\left(\frac{\lambda y + y_1}{\lambda + 1}\right)^2 = 4a\left(\frac{\lambda x + x_1}{\lambda + 1}\right)$$

$$\Rightarrow \quad (\lambda y + y_1)^2 - 4a(\lambda x + x_1)(\lambda + 1) = 0$$

$$\Rightarrow \quad \lambda^2(y^2 - 4ax) + 2\lambda[yy_1 - 2a(x + x_1)] + y_1^2 - 4ax_1 = 0.$$

This equation being quadratic in λ, gives the values of λ which corresponds to points Q and R. For the line to be tangents both the values of λ must be same. Which gives

$$[yy_1 - 2a(x + x_1]^2 = (y^2 - 4ax)(y_1{}^2 - 4ax_1)$$

this is the required equation of the pair of tangents drawn from the point (x_1, y_1) to the parabola $y^2 = 4ax$.

PROPERTIES OF POLE AND POLAR

(1) The chord of a parabola having a given point as its middle point is parallel to the polar of that point.

Let the equation of the parabola be $y^2 = 4ax$ and (h, k) the co-ordinates of the given middle point. Equation of a line through (h, k) is given by

$$y - k = m(x - h) \qquad \text{...(i)}$$

and equation of the polar of (h, k) is given by

$$yk = 2a(x + h). \qquad \text{...(ii)}$$

The points of intersection of line (i) and the parabola is given by

$$y - k = m\left(\frac{y^2}{4a} - h\right)$$

$$\Rightarrow \qquad my^2 - 4ay + 4ak - 4ahm = 0$$

It shows that $y_1 + y_2 = 4a/m$. Since (h, k) is the middle point of the chord, therefore, $\frac{y^2 + y_2}{2} = 2a/m$ or $m = 2a/k$, which is also the slope of the line (ii). Hence the lines (i) and (ii) are parallel.

(2) If the polar of P with respect to a parabola passes through Q, then the polar of Q passes through P.

Let the parabola be $y^2 = 4ax$ and the co-ordinates of P and Q be (x_1, y_1) and (x_2, y_2) respectively.

The polar of P is given by

$$yy_1 = 2a(x + x_1). \qquad \text{...(i)}$$

The polar of Q is given by

$$yy_2 = 2a(x + x_2). \qquad \text{...(ii)}$$

If the polar of P Passes through Q, then we have

$$y_1y_2 = 2a(x_1 + x_2). \qquad \text{...(iii)}$$

Again if the polar of Q passes through P, then we have

$$y_1y_2 = 2a(x_1 + x_2). \qquad \text{...(iv)}$$

From (iii) and (iv), we conclude that if the polar of P passes through Q, then the polar of Q will pass through P. The points P and Q are called *Conjugate Points.*

If the pole of a line $u = 0$ lies on the line $v = 0$, then the pole of $v = 0$ will lie on $u = 0$.

Let (x_1, y_1) and (x_2, y_2) be the poles of $u = 0$ and $v = 0$ then

we have $u \equiv yy_1 - 2a(x + x_1) = 0$ ...(i)

$$v \equiv yy_2 - 2a(x + x_2) = 0 \quad \text{...(ii)}$$

If the pole of the line $u = 0$ lies on $v = 0$, *i.e.*, (x_1, y_1) lies on $v = 0$, then we have

$$y_1y_2 - 2a(x_1 + x_2) = 0. \quad \text{...(iii)}$$

This is also the condition that (x_2, y_2), the pole of the line $v = 0$ lies on $u = 0$. Hence the result. The lines $u = 0$ and $v = 0$ are called *conjugate lines.*

Cor. *Obtain the condition that the two straight lines $l_1x + m_1y + n_1 = 0$ and $l_2x + m_2y + m_2 = 0$ are conjugate with respect to the parabola $y^2 = 4ax$.*

The pole of $l_1x + m_1y + n_1 = 0$ *with respect to parabola* $y^2 = 4ax$ *is*

$$\left(\frac{n_1}{l_1}, \frac{-2am_1}{l_1}\right).$$

This pole will lie on $l_1x + m_1y + n_1 = 0$ if

$$l_1\left(\frac{n_1}{l_1}\right) + m_2\left(\frac{-2am_1}{l_1}\right) + n_2 = 0$$

or $\qquad l_1n_2 + l_2n_1 = 2am_1m_2,$

which is the required condition.

POLE AND POLAR

Definition : *Let P be any point. The polar of P is the locus of the points of intersection of the tangents drawn to the parabola at the ends of chord passing through P.*

TO FIND THE EQUATION OF THE POLAR OF A GIVEN POINT WITH RESPECT TO A PARABOLA

Let $P(x_1, y_1)$ be the given point. Through P a chord is drawn meeting the parabola is Q and R. The tangents at Q and R intersect at T. The locus of T is called the polar of P and P is called the pole. Let the co-ordinates of T be (h, k). Since QR is the chord of contact of T, therefore its equation is given as

$$yk = 2a(x + h). \quad \text{...(i)}$$

As QR passes through P, it gives

$$y_1k = 2a(x_1 + h). \qquad ...(ii)$$

$\therefore$ locus of T is $yy_1 = 2a(x + x_1)$.

This is the required equation of the polar.

POLE OF A STRAIGHT LINE

To find the pole of the line $lx + my + n = 0$ with respect to the parabola $y^2 = 4ax$.

Let (x_1, y_1) be the pole of the line $lx + my + n = 0$ with respect to the parabola $y^2 = 4ax$, then we have

$$yy_1 = 2a(x + x_1) = 0$$

must represent the same straight line as $lx + my + n = 0$.

Comparing the co-efficients, we have

$$\frac{y_1}{m} = \frac{-2a}{l} = \frac{-2ax_1}{n}$$

$$\Rightarrow \qquad x_1 = \frac{n}{l}, \text{ and } y_1 = \frac{-2am}{l}$$

Hence the pole of $lx + my + n = 0$ is

$$\left(\frac{n}{l}, \frac{-2am}{l}\right).$$

CHORD OF CONTACT

To find the equation of the chord of contact of tangents drawn to a parabola from a given point outside it.

Let O(h, k) be any point lying out side the parabola $y^2 = 4ax$. From O two tangents OP and OQ are drawn touching the parabola at the point P and Q.

The equation of the tangents at P and Q are given as

$$yy_1 = 2a(x + x_1) \qquad ...(i)$$

$$yy_2 = 2a(x + x_2) \qquad ...(ii)$$

Since these tangents pass through O(h, k), so we have

$$ky_1 = 2a(h + x_1) \qquad ...(iii)$$

$$ky_2 = 2a(h + x_2). \qquad ...(iv)$$

Equations (iii) and (iv) show that both the points $P(x_1, y_1)$ and $Q(x_2, y_2)$ satisfy the equation

$$ky = 2a(x + h). \qquad ...(v)$$

Hence (v) represents the required equation of the chord of contact.

EQUATION OF A CHORD WHOSE MIDDLE POINT IS GIVEN

Let $P(x_1, y_1)$, $Q(x_2, y_2)$ be the end points of a chord PQ whose middle point is (h, k). Equation of the chord passing through (h, k) is given by

$$y - k \; m(x - h). \qquad ...(i)$$

Slope of the line $PQ = \frac{y_2 - y_1}{x_2 - x_1}$. Since P and Q lie on the parabola $y^2 = 4ax$, so we have

$$y_1^2 = 4ax_1.$$
$$y_2^2 = 4ax_2.$$

Subtraction and simplification gives

$$\frac{y_2 - y_1}{x_2 - x_1} = \frac{4a}{y_1 + y_2}.$$

$\therefore$ Slope of (i) is given by

$$m = \frac{4a}{y_1 + y_2}$$

Since (h, k) is the middle points of PQ, so we get

$$k = \frac{y_1 + y_2}{2}$$

This gives $\qquad m = \frac{4a}{k} = \frac{2a}{k}$

and (i) becomes

$$y - k = \frac{2a}{k}(x - h),$$

which is the required equation of the chord whose middle point is (h, k).

SOLVED EXAMPLES

Example 1:

Find the locus of a point O when the three normals drawn from it are such that the line joining the feet of two of them is always in a given direction.

Solution:

Let $A \equiv \left(am_1^2, -2am_1\right)$ and $B \equiv \left(am_2^2, -2am_2\right)$ be the co-ordinates of the feet of any two of the three normals. Now slope of the line joining AB is

$$= \frac{-2a\left(m_1 - m_2\right)}{a\left(m_1^2 - m_2^2\right)} = \frac{-2}{m_1 + m_2} = \text{constant.[by hypothesis]}$$

$\Rightarrow \quad m_1 + m_2 = \text{constant} = -1 \text{ (say)}.$

As $m_1 + m_2 + m_3 = 0$ by Equation (1) of Q. No. 1.

so $\quad m_3 = -(m_1 + m_2) = \lambda.$

Again, we have

$$m_3 (m_2 + m_1) + m_1 m_2 = \frac{2a-h}{a} \text{ or } -m_3^2 - \frac{k}{a\lambda} = \frac{2a-h}{a}$$

$$\Rightarrow \quad -\lambda^2 - \frac{k}{a\lambda} = \frac{2a-h}{a}$$

$$\Rightarrow \quad al^3 + (2a - h)\lambda + k = 0.$$

Generalising the locus of (h, k) is $a\lambda^3 + (2a - x)\lambda + y = 0$ which is a normal at $(a\lambda^2, -a\lambda)$ to the parabola.

Example 2(a):

Prove that the normals at the points, where the straight line $lx + my = 1$ meets the parabola, meet on the normal at the point

$$\left(\frac{4am^2}{l^2}, \frac{4am}{l}\right)$$

of the parabola.

Solution:

The line is $lx + my = 1$. ...(1)

and the parabola is $y^2 = 4ax$.

Let (1) cut the parabola in P and Q such that

$$P \equiv \left(at_1^2, -2at_1\right) \text{ and } Q \circ \left(at_2^2, -2at_2\right).$$

Now equation of PQ is $(t_1 + t_2)\, y + 2x + 2at_1t_2 = 0$ ...(2)

Since the lines (1) and (2) represent the same line, so comparing the coefficients, we get

$$\frac{t_1 + t_2}{m} = \frac{2}{l} = \frac{2at_1\, t_2}{-1}; \text{ so } t_1 + t_2 = \frac{2m}{l} \qquad ...(3)$$

and $\quad t_1\, t_2 = \dfrac{1}{al}.$ ...(4)

Again, the normal at $(at^2, -2at)$ is $y = tx - 2at - at^3$, or $at^3 + t(2a - x) + y = 0$.

The roots of this equation will be t_1, t_2 and t_3, such that $t_1 + t_2 + t_3 = 0$.

So $\quad t_3 = -(t_1 + t_2) = \dfrac{-2m}{l} \quad$ [by (3)] ...(5)

As the third point on the parabola is $\left(at_3^2, -2at_3\right)$, so putting the value of t_3 from (5), we get the required point as $\left(\frac{4m^2}{l^2}, \frac{4am}{l}\right)$. **Proved.**

Example 2(b):

Show that the locus of the point of intersection of two tangents, which the tangent at the vertex form a triangle of constant area c^2, is the curve $x^2 (y^2 - 4ax) = 4c^4$.

Solution:

Let $P \equiv (h, k)$ be the point of intersection of tangents of the parabola $y^2 = 4ax$, and if these cut an intercept AB on the y-axis, then $AB = \sqrt{(k^2 - 4ah)}$. The length of perpendicular from $P \equiv (h, k)$ on AB *i.e.*, $y = 0$ is clearly h. Hence the area of the triangle ABP

$$= \frac{1}{2}\sqrt{(k^2 - 4ah)}\,.\,h = c^2 \qquad \text{(by hypothesis).}$$

Squaring, simplifying and generalising, the locus of (h, k) is given as

$$x^2 (y^2 - 4ax) = 4c^4.$$ **Proved.**

Example 3:

Two tangents to a parabola meet at angle of 45°; prove that the locus of their point of intersection is the curve $y^2 - 4ax = (x + a)^2$.

If they meet at an angle of 60°, prove that the locus is $y^2 - 3x^2 - 10ax - 3a^2 = 0$.

Solution:

(a) Let m_1 and m_2 be the slopes of two tangents through P.

As the angle between them is 45°, we have

$$\tan 45^\circ = \pm\frac{m_1 - m_2}{1 + m_1 m_2} = \pm\frac{\sqrt{\left\{(m_1 + m_2)^2 - 4\, m_1\, m_2\right\}}}{1 + m_1\, m_2}.$$

Putting the values of m_1 and m_2, we get

$$1 = \pm\frac{\sqrt{\left(\frac{k^2}{h^2} - \frac{4a}{h}\right)}}{\left(1 + \frac{a}{h}\right)}$$

$$\Rightarrow \qquad 1 = \frac{\sqrt{(k^2 - 4ah)}}{h + a}.$$

Generalising after squaring and simplifying, we get the locus of (h, k) as $y^2 - 4ax = (x + a)^2$.

(b) Again, if the angle between the tangents if 60°, then

$$\tan 60° = \pm\frac{m_1 - m_2}{1 + m_1\ m_2} = \pm\frac{\sqrt{\left\{(m_1 + m_2)^2 - 4\ m_1\ m_2\right\}}}{1 + m_1\ m_2}$$

$$\Rightarrow \qquad \sqrt{3} = \pm\frac{\sqrt{(k^2 - 4ah)}}{a + h}, \text{ [as in part (a)].}$$

Squaring, simplifying and generalising, the required locus is

$$y^2 - 3x^2 - 10ax - 3a^2 = 0.$$

Example 4:

From an external point P tangents are drawn to the parabola; find the equation to the locus of P when these tangents make angles q_1 and q_2 with the axis, such that tan q_1 + tan q_2 is constant (= b).

Solution:

Let the co-ordinates of P be (h, k) and the equation to the parabola be $y^2 = 4ax$.

Any tangent on the parabola is given by y = mx + a/m. If this passes through (h, k), the co-ordinates will satisfy. Hence k = mh + a/m

$$\Rightarrow \qquad m^2h - mk + a = 0. \qquad ...(1)$$

[1] is a quadratic in m. Let its roots be m_1 and m_2, then $m_1 + m_2 = k/h$ and $m_1\ m_2 = a/h$. Now, if the two tangents through P make angles θ_1 and θ_2 with axis of x_1 then $m_1 = \tan\theta_1$ and $m_2 = \tan\theta_2$.

$$\therefore \qquad \tan\theta_1 + \tan\theta_2 = k/h. \qquad ...(2)$$

$$m_1\ m_2 = a/h \qquad ...(3)$$

$$\Rightarrow \qquad \tan\theta_1\ \tan\theta_2 = a/h \qquad ...(4)$$

By hypothesis, $\tan\theta_1 + \tan\theta_2 = b$

So from (2), k/h = b or k = bh

Generalising, the locus of (h, k) is y = bx. **Ans.**

Example 5:

Prove that the locus of the centre of a circle, which intercepts a chord of given length 2a on the axis of x and passes through a given point on the axis of y distant b from the origin, is the curve

$$x^2 - 2yb + b^2 = a^2.$$

Trace this parabola.

Solution:

Let the centre of the required circle be (h, k); so its equation can be written as $x^2 + y^2 - 2hx - 2ky + c = 0$. ...(1)

The circle passes through a point on y-axis which is at a distance of be from the origin The co-ordinates of the point will be (0, b). These co-ordinates will satisfy (1); hence

$$0 + b^2 - 0 - 2kb + c = 0$$

$$\Rightarrow \quad b^2 - 2kb + c = 0. \qquad ...(2)$$

Again, solving (1) with x-axis *i.e.* y = 0, we get

$$x^2 - 2hx + c = 0 \qquad ...(3)$$

If the circle cuts x-axis at A and B, the abscissae of A and B say x_1 and x_2 are given by (3). The distance AB will be $(x_1 - x_2) = \sqrt{\{(x_1 - x_2)^2 - 4x_1 x_2\}}$ = 2a (by hypothesis).

By (3), $x_1 + x_2 = 2h$ and $x_1x_2 = c$. So $2a = \sqrt{\{2h)^2 - 4c\}}$

or $\quad 4a^2 = 4h^2 - 4c \quad$ or $\quad h^2 - c - a^2 = 0$ ...(4)

The required locus will be obtained by eliminating c from (2) and (4).

So adding these two, we get $b^2 - 2kb - h^2 - a^2 = 0$.

Generalising, the required locus is

$$x^2 - 2yb + b^2 - a^2 = 0.$$ **Proved.**

This can be written as $x^2 = 2yb + a^2 - b^2$

or $x^2 = 2b\left(y + \dfrac{a^2 - b^2}{2b}\right)$ ...(5)

Changing the origin to $\left(0, -\dfrac{a^2 - b^2}{2a}\right)$ the curve becomes

$$X^2 = 2bY \qquad ...(6)$$

Where X and Y are new variables. The vertex of the parabola (6) is (0, 0) focus (0, – b/2), axis is X = 0 *i.e.* Y-axis and latus rectum is 2b. Transferring to the original axis, the co-ordinates of the vertex will be $\left(0, -\dfrac{a^2 - b^2}{2a}\right)$ axis of the parabola will be the same x-axis and latus rectum will be 2b; hence the curve may be drawn.

Example 6:

Find the vertex, axis, latus rectum and focus of the parabola

$$x^2 + 2y = 8x - 7$$

Solution:

The given equation is

$$x^2 + 2y = 8x - 7$$

or $\quad x^2 - 8x = -2y - 7$

$\Rightarrow \quad x^2 + 8x + 16 = -2y - 7 + 16$

or $\quad (x - 4)^2 = -2y + 9$

$\Rightarrow \quad (x - 4)^2 = -2\left(y - \dfrac{9}{2}\right)$

or $\quad (x - 4)^2 = -4\,\dfrac{1}{2}\cdot\left(y - \dfrac{9}{2}\right) \qquad ...(1)$

Changing the origin to (4, 9/2), the equation (1) becomes

$$X^2 = -4.\ \frac{1}{2}.\ Y. \qquad ...(2)$$

where X and Y are new variables.

According to t he new axis, the vertex of (2) is (0, 0), its axis Y axis or X = 0; latus rectum is 4.(1/2) = 2, and the focus is (0, – 1/2) (as a = – 1/2).

Changing back to the original axes the vertex will be (4, 9/2), axis will be X = 0 *i.e.* x – 4 = 0; latus rectum is 2 and the focus is (0 + 4, – 1/2 = 9/2) *i.e.* (4, 4) **Ans.**

Example 7(a):

Find the value of p when the parabola $y^2 = 4px$ goes through the point (i) (3, – 2) and (ii) (9, – 12).

Solution:

The curve is $\quad y^2 = 2px. \qquad ...(1)$

(i) If the curve passes through (3, – 2), the co-ordinates will satisfy (1);

so $4 = 4p.\ 3$ or $p = 1/3$ **Ans.**

(ii) If the curve passes through (9, – 12), the co-ordinates will satisfy (1);
r $144 = 4p.\ 9$ or $p = 4$. **Ans.**

Example 7(b):

Show that the locus of the middle point of chords of the parabola $y^2 = 4ax$, which are of constant length 2l is

$$(4ax - y^2)(y^2 + 4a^2) = ,4a^2l^2.$$

Solution:

Let $R(x_1, y_1)$ be the middle point of the chords PQ of the parabola $y^2 = 4ax$. We also assume that the co-ordinates of P and Q are $(at_1^2, 2at_1)$ and $(at_2^2, 2at_2)$.

Since the length of the chord is $2l$, it gives

$$\sqrt{[(at_1^2 - at_2^2)^2 + (2at_1 - 2at_2)^2]} = 2l$$

$$\Rightarrow \quad a^2(t_1^2 - t_2^2)^2 + 4a^2(t_1 - t_2)^2 = 4al^2 \quad ...(i)$$

Also $$x_1 = \frac{at_1^2 + at_2^2}{2} \Rightarrow t_1^2 + t_2^2 = 2x_1/a \quad ...(ii)$$

$$y_1 = \frac{2a(t_1 + t_2)}{2} \Rightarrow t_1 + t_2 = y_1/a. \quad ...(iii)$$

Eliminating t_1, t_2 from equations (i), (ii) and (iii), we have

$$(4x_1a - y_1^2)(y_1^2 + 4a^2) = 4l^2a^2.$$

∴ Locus is

$$(xa - y^2)(y^2 + 4a^2) = 4a^2l^2.$$

Example 8:

A tangent to the parabola $y^2 + 4bx = 0$ meets the parabola $y^2 = 4ax$ at P and Q. Prove that the locus of the middle point of PQ is $y^2(2a + b) = 4a^2x$.

Solution:

Let the mid-point of PQ be (x_1, y_1). PQ being a chord of $y^2 = 4ax$ having (x_1, y_1) as its mid-point, its equation as

$$yy_1 - 2a(x + x_1) = y_1^2 - 4ax_1 \qquad (S_1 = T)$$

$$\Rightarrow \quad y = \frac{2ax}{y_1} + \frac{y_1^2 - 2ax_1}{y_1} \quad ...(i)$$

The equation (i) will be tangent to the parabola $y^2 = -4bx$

if $\frac{y_1^2 - 2ax_1}{y_1} = \frac{-b}{2a/y_1}$ $\left(c = \frac{a}{m}\right)$

$\Rightarrow$ $y_1^2 - 2ax_1 + - (b/2a)\, y_1^2$

$\Rightarrow$ $y_1^2 (2a + b) = 4a^2x_1.$

Hence the locus of (x_1, y_1) is

$$y^2 (2a + b) = 4a^2x.$$

Example 9:

Show that the locus of the poles of the normal chords of the parabola $y^2 = 4ax$ is

$$(x + 2a)\, y^2 + 4a^3 = 0.$$

Solution:

We know that the equation of normal in the slope form is

$$y = mx - 2am - am^3 \quad \text{...(i)}$$

Let its pole be (h, k).

The equation of the polar of (h, k) with respect to the parabola $y^2 = 4ax$ is

$$yk = 2a(x + h) \quad \text{...(ii)}$$

$\therefore$ (i) and (ii) are identical, so we get after comparing coefficients

$$\frac{k}{1} = \frac{2a}{m} = \frac{2ah}{-2am - am^3}.$$

This gives $m = \frac{2a}{k}$

and $2am + am^3 = \frac{-2ah}{k}$

$\Rightarrow$ $2a \cdot \frac{2a}{k} + a\left(\frac{2a}{k}\right)^3 = \frac{-2ah}{k}$

$\Rightarrow$ $4a^2k^2 + 8a^2 = -\, 2ahk^2.$

$\Rightarrow$ $(h + 2a)\, k^2 + 4a^3 = 0.$

$\therefore$ locus of (h, k) is

$$(x + 2a)\, y^2 + 4a^3 = 0.$$

Example 10:

Prove that the equation $y^2 + 2Ax + 2By + C = 0$ represents a parabola, whose axis is parallel to the axis of a, and find its vertex and the equation to its latus rectum.

Solution:

The given equation is $y^2 + 2Ax + 2By + C = 0$

$\Rightarrow \quad y^2 + 2By + 2Ax - C$

$$\Rightarrow \quad (y + B)^2 = -2Ax - C + B^2 = -2A\left(x - \frac{B^2 - C}{2A}\right) \quad ...(1)$$

Transferring the origin at $\left(\frac{B^2 - C}{2A}, -B\right)$ the equation (1) becomes

$$Y^2 = -2AX. \quad ...(2)$$

where X and Y are new variables.

For equation No. (2), the vertex is (0, 0) axis of parabola is X-axis or Y = 0, and the focus is (– A/2, 0); hence the equation to latus rectum is X = – A/2.

Changing the origin back to the initial one, the co-ordinates of the vertex according to the original axes will be

$$\left(\frac{B^2 - C}{2A}, -B\right) \quad \textbf{Ans.}$$

The axis of the parabola will be y = – B or y + B = 0 which is parallel to axis of X.

Also, the equation to latus rectum will be

$$x = \frac{B^2 - C}{2A} - \frac{A}{2}$$

or $$x = \frac{B^2 - A^2 - C}{2A} \quad \textbf{Ans.}$$

Example 11:

Prove that the equation to the parabola whose vertex and focus are on the axis of x at distances a and a' from the origin respectively, is

$$y^2 = 4\,(a' - a)\,(x - a).$$

Solution:

By hypothesis vertex is (a, 0) and focus is (a', 0). The directrix is at a distance equal to the distance of focus and on the opposite side of the vertex. The distance between the vertex and focus is (a' – a) = (2a – a'). As the directrix is perpendicular to the axis of parabola and here the axis is same as axis of x (as the line joining the vertex and focus is x-axis), the directrix will be parallel to y-axis. So its equation will be x = 2a – a'.

Let there be any point $P \equiv (h, k)$ on the parabola.

Distance of P from the directrix $= h - (2a - a')$. ...(1)

Distance between $P \equiv (h, k)$ and the focus $(a, 0)$

$$= \sqrt{\{(h - a)^2 + (k - 0)^2\}}.$$

As the curve is a parabola, by the definition,

$$\{h - (2a - a')\} = \sqrt{\{(h - a)^2 + k^2\}}.$$

Squaring and simplifying, we get

$$k^2 = 4a'h + 4a^2 - 2ha - 4aa'$$

$$\Rightarrow \quad k^2 = 4\,[(a' - a)\,(h - a)].$$

Generalising, the required curve is

$$y^2 = 4\,(a' - a)\,(x - a).$$ **Proved.**

Example 12(a):

PQ is a double ordinate of a parabola. Find the locus of its points of trisection.

Solution:

Let the equation to the parabola be $y^2 = 4ax$.

If x_1 is the abscissa of any point, the value of the ordinate will be $\pm 2\sqrt{(ax_1)}$. Hence the end points of the double ordinate are $\{a_1, 2\sqrt{(ax_1)}\}$ and $\{a_1 - 2\sqrt{(ax_1)}\}$.

Say (h, k) are the co-ordinates of the point dividing the double ordinate in the ration of 1:2, then

$$h = \frac{x_1 + 2x_1}{3} = x_1 \qquad ...(1)$$

and

$$k = \frac{-2\sqrt{(ax_1)} + 2.2\sqrt{(ax_1)}}{3} = \frac{2\sqrt{(ax_1)}}{3}. \qquad ...(2)$$

The required curve is the eliminant of x_1 from (1) and (2).

Hence squaring (2) and putting the value of x_1 from (1), we get

$$k^2 = \frac{4}{9}a.h$$

Generalising, we get the required locus as

$$y^2 = \frac{4}{9}ax$$

or $9y^2 = 4\,ax$. **Ans.**

Example 12(b):

At the ends of the latus rectum of the parabola $y^2 = 12x$.

Solution:

Parabola is $y^2 = 12x = 4.\ 3x$. Hence $a = 3$.

As the ends of the latus rectum are (a, 2a) and (a, – 2a), the points here will be (3, 6) and (3, – 6).

Equation of the tangent at (3, – 6) to parabola $y^2 = 12x$ is

$$y\ .\ 6 = 6\ (x + 3) \quad \Rightarrow \quad y - x = 3.$$ **Ans.**

Normal at (3, 6) will be

$$y - 6 = -\frac{6}{6}\ (x - 3) \quad \Rightarrow \quad y + x = 9.$$ **Ans.**

Again, equation of the tangent at (3, – 6) will be

$$y\ (-6) = 6\ (x + 3) \quad \Rightarrow \quad x + y + 3 = 0$$ **Ans.**

and the normal at (3, – 6) will be

$$y - (-6) = -\frac{-6}{6}\ (x - 3) \quad \Rightarrow \quad x - y = 9.$$ **Ans.**

Example 13:

A parabola is drawn to pass through A and B, the ands of a diameter of a given circle of radius a, and to have as directrix a tangent to a concentric circle of radius b; the axes being AB and a perpendicular diameter, prove that the locus of the focus of the parabola is

$$\frac{x^2}{b^2} + \frac{y^2}{b^2 - a^2} = 1.$$

Solution:

Take origin at the centre of the circle, x-axis along the diameter AB and y-axis along the perpendicular to AD through the centre.

As the radius of the first circle is given as a, the co-ordinates of A and B may be taken as (– a, 0) and (a, 0) respectively.

Centre to any concentric circle to this will be again the origin. If the radius of the circle be b, its equation will be

$$x^2 + y^2 = b^2. \qquad ...(1)$$

Equation of any tangent to (1) is given as

$$y = mx + b\ \{\sqrt{(1 + m^2)}\} \qquad ...(2)$$

Hence the directrix given by (2).

Let (h, k) be the co-ordinates of the focus S.

If A lies on the parabola. Its distance from the focus and the directrix must be the same; so we have

$$\sqrt{\left\{(h+a)^2+k^2\right\}} = \frac{-ma+b\sqrt{\left(1+m^2\right)}}{\sqrt{\left(1+m^2\right)}}$$

$\Rightarrow \sqrt{(1 + m^2)}\ \sqrt{(h + a)^2 + k^2\}} = -\ ma + b\sqrt{(1 + m^2)}$ ...(3)

Similarly as B lies on the parabola, we will get

$\sqrt{(1 + m^2)}\ \sqrt{(H + a)^2 + k^2\}} = ma + b\sqrt{(1 + m^2)}$ ...(4)

The reqd. locus will be obtained by eliminating m from (3) and (4). Adding (3) and (4), we get

$\sqrt{(1 + m^2)}\ [\sqrt{\{(h - a)^2 + k^2\}} + \sqrt{\{(h + a)^2 + k^2\}}] = 2b\sqrt{(1 + m^2)}$

$\Rightarrow \sqrt{\{(h - a)^2 + k^2\}} = 2b - \sqrt{\{(h + a)^2 + k^2\}}$

Squaring, we get

$h^2 + a^2 - 2ah + k^2 = 4b^2 + h^2 + a^2 + 2ah\ k^2 - 4b\ \sqrt{\{(h + a)^2 + k^2\}}$

$\Rightarrow - 4b^2 - 4ah + - 4b\ \sqrt{\{(h + a)^2 + k^2)}$

Dividing by 4 as it is a common factor and again squaring,

$b^4 + a^2h^2 + 2ab^2h = b^2\ (h^2 + a^2 + 2ah + k^2)$

$\Rightarrow h^2\ (b^2 + a^2) + k^2b^2 = (b^4 - b^2a^2)$

$\Rightarrow h^2\ (b^2 - a^2) + k^2b^2 = b^2\ (b^2 - a^2).$

Dividing by R. H. S. and generalising, the reqd, locus is

$$\frac{x^2}{b^2}+\frac{y^2}{b^2-a^2}=1$$ **Ans.**

Example 14:

Through a point P are drawn tangents PQ and PR to a parabola and circles are drawn through the focus to touch the parabola in Q and R respectively; prove that the common chord of these circles passes through the centroid of the triangle PQR.

Solution:

If the co-ordinates of Q and R be $\left(at_1^2, 2t_1\right)$ and $\left(at_2^2, 2t_2\right)$ respectively, then

the equation of tangent at Q is $t_1y = x + at_1^2$,

the equation of the tangent at R is $t_2y = x + at_2^2$.

So the co-ordinates of the point of intersection P will be $\{at_1t_2 . a(t_1 + t_2)\}$. Let the co-ordinates of the centroid of the ΔPQR be (h, k).

Then $h = \frac{a}{3}\left(t_1^2 + t_2^2 + t_1 t_2\right)$ and $k = \frac{a}{3}\left(2t_1 + 2t_2 + t_1 + t_2\right)$

or $\quad k = a\ (t_1 + t_2)$

the equation of the circle which passes through focus (a, 0) and touches the parabola at point $\left(at_1^2, 2at_1\right)$ is

$$x^2 + y^2 - ax\left(3t_1^2 + 1\right) - ay\left(3t_1 - t_1^3\right) + 3a^2t^2 = 0. \quad ...(1)$$

Similarly other circle touching the parabola at $\left(at_2^2, 2at_2\right)$ is

$$x^2 + y^2 - ax\left(3t_2^2 + 1\right) - ay\left(3t_2 - t_2^3\right) + 3a^2t_2^2 = 0 \quad ...(2)$$

Now the common chord of the circles is obtained by subtracting (2) from (1), so its equation is

$$-ax\left(3t_1^2 - 3t_2^2\right) - ay\left(3t_1 - 3t_2 - t_1^3 + t_2^3\right) + 3a^2\left(t_1^2 - t_2^2\right) = 0$$

$\Rightarrow \quad 3ax\ 3ax\left(t_1 + t_2\right) - ay\left(t_1^2 + t_2^2 + t_1t_2 - 3\right) = 3a^2\ (t_1 + t_2)$

$\Rightarrow \quad kx - y\ (h - a) = ak.$

This line is satisfied by (h, k). **Hence proved.**

Example 15:

The sum of the intercepts which the normals cut off from the axis is 2 (h + a).

Solution:

Let P, Q and R be the three points on the parabola having the co-ordinates as $\left(am_1^2, -2am_1\right)$, $\left(am_2^2, -2am_2\right)$ and $\left(am_3^2, -2am_3\right)$ respectively. Then equations of normal at these points are respectively

$$y = m_1x - 2am_1 - am_1^3, \quad ..(1)$$

$$y = m_2x - 2am_2 - am_2^3 \quad ...(2)$$

and $\quad y = m_3x - 2am_3 - am_3^3. \quad ...(3)$

Solving these with the axis of the parabola *i.e.*, y = 0, we get the intercepts on the x-axis respectively as $2a + am_1^2$, $2a + am_2^2$ and $2a + am_3^2$.

Sum of the intercepts made by the normals on the axis

$$= \left(2a + am_1^2\right) + \left(2a + m_2^2\right) + \left(2a + am_3^2\right)$$

$$= 6a + a\ \left(m_1^2 + m_2^2 + m_3^2\right)$$

$$= 6a + a\ \{(m_1 + m_2 + m_3)^2 - 2\ (m_1m_2 + m_2m_3 + m_3m_1)\}$$

$$= 6a + a\ \left\{0 - \frac{2\ (2a - h)}{a}\right\} = 2h + 2a = 2\ (h + a).$$

Example 16:

The normals at three points P, Q and R of the parabola $y^2 = 4ax$ meet in a point O whose coordinates are h and k; prove that the centroid of the triangle POR lies on the axis.

Solution:

Let the co-ordinates of P, Q and R be $\left(am_1^2, -2am_1\right)$, $\left(am_2^2, -2am_3\right)$ and $\left(am_3^2, -2am_3\right)$ respectively. Now, if O ≡ (x, y) be the co-ordinates of the centroid of ΔPQR, then we have

$$y = \frac{-2am_1 - 2am_2 - 2am_3}{3} = \frac{-2a\ \left(m_1 + m_2 + m_3\right)}{3} = 0$$

Hence the locus of the centroid is y = 0 or x-axis which is axis of the parabola. **Proved.**

Example 17:

The sum of the squares of the sides of the triangle PQR is equal to 2 (h – 2a) (h + 10a).

Solution:

Let the co-ordinates of P, Q and R be $\left(am_1^2, -2am_1\right)$, $\left(am_2^2, -2am_2\right)$ and $\left(am_3^2, -2am_3\right)$ respectively.

Then $PQ^2 = a^2\ \left(m_1^2 - m_2^2\right)^2 + (2am_1 - 2am_2)^2$

$$= a^2\ \left(m_1^2 - m_2^2\right)^2 + 4a^2\ (m_1 - m_2)^2$$

$$= a^2\ (m_1 - m_2)^2\ \{(m_1 + m_2)^2 + 4\}$$

$$= a^2\ \left(m_1^2 - m_2^2 - 2m_1m_2\right)\left(m_3^2 + 4\right)$$

(as $m_1 + m_2 + m_3 = 0$)

$$= a^2\left\{\left(m_1^2m_3^2 + m_2^2m_3^2\right) + 4\left(m_1^2 + m_2^2\right) - 2m_1m_2m_3^2 - 8m_1m_2\right\}$$

Similarly , we can write the value of PR^2 and QQ^2.

Hence $PQ^2 + QR^2 + PR^2$

$$= a^2\left(2\Sigma m_1^2m_2^2 - 8\Sigma m_1^2 + 2m_1m_2m_3\Sigma m_1 - 8\Sigma m_1m_2\right)$$

Now, $Sm_1 = 0$, we have

$$\Sigma m_1^2 = (\Sigma m_1)^2 - 2\Sigma m_1 m_2 = -2\left(\frac{2a-h}{a}\right)$$

and $$\Sigma m_1^2 m_2^2 = (\Sigma m_1 m_2)^2 - (2m_1 m_2 m_3 \Sigma m_1) = \left(\frac{2a-h}{a}\right)^2$$

$$a^2\left[2\left(\frac{2a-h}{a}\right) + 8(-2)\left(\frac{2a-h}{a}\right) - 8\left(\frac{2a-h}{a}\right)\right]$$

$$= 2a^2\left(\frac{2a-h}{a}\right)\left[\frac{2ah}{a} - 12\right] = 2(h-2a)(h+10a)$$

Example 18(a):

If a circle be drawn so as always to touch a given straight line and also a given circle, prove that the locus of its centre is a parabola.

Solution:

As in the last question, taking the centre of the fixed circle as origin, the axes being perpendicular and parallel to the given line respectively, the equation of the given circle will be

$$x^2 + y^2 = r^2 \quad ...(1)$$

and of the line $x = a$ ...(2)

where r is the radius of the circle and a is the distance of the centre of the given circle from the line.

Let the centre of the other circle and b (h, k). So its equation may be written as $x^2 = y^2 - 2hx - 2ky + c = 0$ ...(3)

Radius of (3) is $\sqrt{(h^2 + k^2 - c)}$.

If (1) touches (3), the distance between the centres must be equal to the sum of the radii. So

$$\sqrt{(h^2 + k^2)} = \sqrt{(h^2 + k^2 - c)} = r. \quad ...(4)$$

If (3) touches (2), length of the perpendicular from (h, k) on (2) must be equal to the radius

i.e. $$h - a = \sqrt{(h^2 + k^2 - c)} = r. \quad ...(5)$$

The required locus will be obtained by eliminating c from (4) and (5). So putting the value of $\sqrt{(h^2 + k^2 - c)}$ from (5) in (4), we get

$$\sqrt{(h^2 + k^2)} = (h - a) + r$$

Squaring both sides, $h^2 + k^2 = h^2 = a^2 + r^2 - 2ah + 2rh - 2ar$.

Generalising and simplifying, we get the required locus as

$$y^2 = 2x\,(r - a) + (a - r)^2$$

which is clearly a parabola. **Hence proved.**

Example 18(b):

Prove that the locus of a point, which moves so that its distance from a fixed line is equal to the length of the tangent drawn from it to a given circle, is a parabola. Find the position of the focus and directrix.

Solution:

Let AB be the fixed line, and O be the centre of the fixed circle. Taking O as the origin, a line through O and parallel to AB as axis of y and a line through O perpendicular to AB as axis, the equation of the circle may be written as

$$x^2 + y^2 = r^2 \quad ...(1)$$

and the equation of the line as $x = a$. ...(2)

(where the radius of the circle is r and the distance of the centre from the line AB is a).

Let P be any point (h, k) on the locus, PT be the tangent from P to (1) and PN be perpendicular from P upon AB. Then

$$PT = \sqrt{(h^2 + k^2 - r^2)} \text{ and } PN = h - a$$

By hypotesis PT = PT or $PT^2 = PN^2$.

Simplifying and generalising, we get

$$y^2 = -\,2ax + a^2 + r^2, \quad ...(3)$$

which is clearly a parabola. **Proved.**

Again,

(3) may be written as $y^2 = -2a\left(x \dfrac{a^2 + r^2}{2a}\right)$

Clearly the focus is $\left\{\left(\dfrac{a^2 + r^2}{2a} - \dfrac{a}{2}\right), 0\right\}$ or $\left(\dfrac{r^2}{2a}, 0\right)$

and the directrix will be $x = \frac{a^2 + r^2}{2a} + \frac{a}{2}$

$\Rightarrow \qquad 2ax = 2a^2 + r^2.$ **Ans.**

Example 19:

A double ordinate of the curve $y^2 = 4px$ is of length 8p; prove that the lines from the vertex to its two ends are at right angles.

Solution:

The parabola is given as $y^2 = 4px$. ...(1)

The double ordinate is 8p; so ordinates are ± 4p.

Putting the value in (1), the corresponding abscissa is 4p.

Hence the co-ordinates of the ends of the given double ordinate are (4p, 4p) and (4p, – 4p). Let these points be A and B respectively and the vertex be O ≡ (0, 0).

Slope of OA $= \frac{4p - 0}{4p - 0} = 1\ m_1$ (say).

Slope of OB $= \frac{-4p}{4p} = -1\ m_2$ (say).

Clearly $m_1 \times m_2 = -1$; so OA and OB are at right angles. **Proved.**

Example 20:

Two parabolas have a common axis and concavities in opposite directions; if any line parallel to the common axis meet the parabolas in P and P', prove that the locus of the middle point of PP' is another parabola, provided that the later recta of the given parabolas are unequal.

Solution:

Let the equation of a parabola by $y^2 = 4ax$. ...(1)

As the other parabola has an opposite concavity, it will be on negative side, and as its latus rectum is different, say b, its equation may be written as

$$y^2 = -4bx. \qquad ...(2)$$

The common axis is x-axis. Let any line parallel to x-axis be

$$y = \lambda. \qquad ...(3)$$

Solving (1) with (3), the point of intersection, say P, is $\left(\frac{\lambda^2}{4a}, \lambda\right)$.

Solving (2) and (3), the point of intersection, say Q, is $\left(-\dfrac{\lambda^2}{4b}, \lambda\right)$.

If (h, k) be the mid point of PQ then $h=\dfrac{1}{2}\left(\dfrac{\lambda^2}{4a}-\dfrac{\lambda^2}{4b}\right)$ ...(4)

and $k=\dfrac{1}{2}(\lambda+\lambda)=\lambda$. ...(5)

Putting the value of λ from (5), in (4)

$$h=\frac{1}{2}\left(\frac{k^2}{4a}-\frac{k^2}{4b}\right)=\frac{k^2}{8}\left(\frac{b-a}{ab}\right)$$

Generalising and simplifying, the reqd. locus is

$y^2=\dfrac{8ab}{b-a}x$ which is a parabola. **Proved.**

If the latera recta are the same clearly h = 0 (by 4). So k = 1 i.e, y = 1 will be the reqd locus

Example 21:

A tangent to the parabola $y^2 = 8x$ makes an angle of 45° with the straight line y = 3x + 5. Find its equation and its point of contact.

Solution:

The parabola is given as $y^2 = 8x$. ...(1)

here 4a = 8 or a = 2

Equation to any tangent to (1) is $y = mx + \dfrac{2}{m}$ ...(2)

The given line is y = 3x + 5. ...(3)

If the angle between (2) and (3) is 45°, clearly

$\tan 45^\circ = \dfrac{m-3}{1+m.3} = 1$ as tan 45° = 1.

or m – 3 = 1 + 3m, where m = – 2.

Putting in (2), the equation to the tangent becomes

$y=-2x+\dfrac{2}{-2}$ or y + 2x + 1 = 0. **Ans.**

As the point of contact is $\left(\dfrac{a}{m^2}, \dfrac{2a}{m}\right)$, it will be

$$\left(\frac{2}{4}, \frac{2.2}{-2}\right) \quad \text{or} \quad \left(\frac{1}{2}, -2\right)$$ **Ans.**

Again, if the angle between (2) and (3) is – 45°, we have

$$\frac{m-3}{1+m.3} = \tan(-45^\circ) = -1 \text{ whence } m = \frac{1}{2}.$$

Putting in (2), the equation to the tangent becomes

$$y = \frac{1}{2}x + \frac{2}{1/2} \quad \text{or} \quad 2y = x + 8$$ **Ans.**

and the corresponding point of contact will be

$$\left(\frac{2}{(1/2)^2}, \frac{2.2}{1/2}\right) \quad \text{or} \quad (8, 8).$$ **Ans.**

Example 22:

If on a given base triangles be described such that the sum of the tangents of the base angles is constant, prove that the locus of the vertices is a parabola.

Solution:

Taking the x-axis along the base and the y-axis as the right bisector of the base, the co-ordinates of the end points of the base, BC of a triangle ABC may be written as (– a, 0) and (a, 0) where BC = 2a. Let the co-ordinates of the vertex A be (a, k).

Slope of the line $AB = \frac{k-0}{h+a}$ = tan B where A, B and C are the angles of the triangle (as base coincide with x-axis).

Again slope of $AC = \frac{k-0}{h-a} = \tan(\pi - C), \tan C = \frac{k}{a-h}$

By hypothesis, tan B + tan C = λ (constt)

$$\Rightarrow \qquad \frac{k}{h+a} + \frac{k}{a-h} = \lambda$$

$$\Rightarrow \qquad ka - kh + kh + ka + \lambda(a^2 - h^2).$$

Generalising and re-arranging, the reqd, locus is $\lambda x^2 = \lambda a^2 - 2ay$, which is parabola.

Example 23:

Write down the equations to the tangent and normal, At the point (4, 6) of the parabola $y^2 = 9x$.

Solution:

The parabola is $y^2 = 9x$ and the point is (4, 6). So the equation to the tangent will be

$$y.6 = \frac{9}{2}(x + 4) \quad \text{or} \quad 4y = 3x + 12. \qquad \textbf{Ans.}$$

Equation to the normal will be

$$y - 6 = -\frac{6}{9/2}(x - 4) \quad \text{or } 3y = 4x + 34. \qquad \textbf{Ans.}$$

Example 24:

At the point of the parabola $y^2 = 6x$ whose ordinate is 12.

Solution:

The parabola is $y^2 = 6x$. If the ordinate of the same point on the parabola is 12, corresponding abscissa will be given by $(12)^2 = 6x$ or $x = 24$. Hence the point is (24, 12)

$$y.12 = 3(x + 24) \quad \text{or} \quad 4y - x = 24. \qquad \textbf{Ans.}$$

Equation of the normal at (24, 12) is

$$y - 12 = -\frac{12}{3}(x - 24) \quad \text{or} \quad y + 4x = 108. \qquad \textbf{Ans.}$$

Example 25:

The vertex A of a parabola is joined to any point P on the curve and PQ is drawn at right angles to AP to meet the axis in Q. Prove that the projection of PQ on the axis the always equal to the latus rectum.

Solution:

Let the equation to the parabola be

$$y^2 = 4ax. \qquad ...(1)$$

If the co-ordinates of any point P on the parabola be (h, k) then

$$k^2 = 4ah. \qquad ...(2)$$

Slope of AP is k/h where A is the vertex (0, 0).

The equation of a line PQ which is perpendicular to AP will be

$$y - k = -\frac{k^2}{k}(x - h)$$

or $\quad yk + hx = k^2 + h^2. \qquad ...(3)$

If Q lies on X-axis, the its ordinate is 0. So putting $y = 0$ is (3),

$$x=\frac{k^2}{h}+h.$$

Putting the value of k^2 from (2),

$$x=\frac{4ah}{h}+h,\ h = (4a + h).$$

The projection of PQ on axis of x is the difference between the abscissae of the points P and Q. Hence the projection

$$= (4a + h) - h = 4a$$

which is equal to the latus rectum. **Proved.**

Example 26:

Find the equation of a parabola. Whose focus (a, b) and directrix $\frac{x}{a}+\frac{y}{b}=1.$

Solution:

The focus S is (a, b) and the directrix AB is

$$\frac{x}{a}+\frac{y}{b}=1 \quad \text{or} \quad bx + ay - ab = 0 \qquad ...(1)$$

If the moving point P be (h, k), PN be the perpendicular from P on AB, then by the property of parabola,

$$PS = PN \quad \text{or} \quad PS^2 = PN^2.$$

So $\quad (h - a)^2 + (k = b)^2 = \frac{(bh+ak-ab)^2}{b^2+a^2}$

$\Rightarrow \quad (a^2 + b^2)(h^2 + a^2 - 2ah + k^2 - 2bk + b^2)$

$= b^2h^2 + a^2k^2 + a^2b^2 + 2abhk - 2ab^2h - 2a^2bk$

$= h^2 (a^2 + b^2 + a^2) k^2 (a^2 + b^2 - a^2) - 2abhk - 2ah (a^2 + b^2 - b^2)$
$- 2bk (a^2 + b^2 - a^2) + a^2 + a^2b^2 + b^4 - a^2b^2 = 0$

Simplifying and generalising, we get

$$a^2x^2 + b^2y^2 + 2abxy - 2a^3x - 2b^3y + a^4 + a^2b^2 + b^4 = 0$$

$\Rightarrow \quad (ax - by)^2 - 2a^3x - 2b^3y + a^4 + a^2b^2 + b^4 = 0.$ **Ans.**

Example 27:

Find the equation to the parabola with

Focus (3, – 4) and directrix $6x - 7y + 5 = 0$.

Solution:

The focus, say S, is given as (3, – 4) and the directrix say AB is

$$6x - 7y + 5 = 0.$$

Let the co-ordinates of the moving point P be (h, k).

If PN be the perpendicular from P to AB, then

$$\text{m PN} = \frac{6h-7k+5}{\sqrt{(6^2+7^2)}} \quad \text{and} \quad PS = \sqrt{\{(h-3)^2+(k+4)^2\}}$$

If the locus is a parabola, PS = PN or $PS^2 = PN^2$

$$\Rightarrow \quad (h-3)^2 + (k+4)^2 = \frac{(6h-7k+5)^2}{36+49}$$

$$\Rightarrow \quad 85\,[h^2 - 6h + 9 + k^2 + 8k + 16]$$

$$= [36h^2 + 49k^2 + 25 - 84hk + 60h = 70k]$$

$$\Rightarrow \quad 36h^2 + 36k^2 + 84hk - 570h + 750k + 2100 = 0$$

$$\Rightarrow \quad (7h + 6k)^2 - 570h + 750k + 2100 = 0.$$

Generalising, we get

$$(7x + 6y)^2 - 570h + 750k + 2100 = 0.$$ **Ans.**

Example 28(a):

Prove that the locus of the middle points of all chords the parabola $y^2 = 4ax$ which are drawn through the vertex is the parabola y^2 $2ax$.

Solution:

The equation to the parabola is given as

$$y^2 = 4ax. \qquad ...(1)$$

Any line passing through the origin may be given by

$$y = mx. \qquad ...(2)$$

Solving (1) and (2) $m^2x^2 = 4ax$

or $\quad m^2x^2 - 4ax = 0$

or $\quad x\,(m^2x - 4a) = 0$

whence either x = 0 or $x = (4a/m^2)$. The corresponding values of y are 0 and 4a/m. Hence (1) and (2) meet in O and A at (0, 0) and $(4a/m^2, 4a/m)$ resp. If (h, k) be the mid-point of this chord OA,

$$h = \frac{0+4a/m^2}{2} = \frac{2a}{2}. \qquad ...(3)$$

and $$k = \frac{0 + 4a/m}{2} = \frac{2a}{m}. \qquad ...(4)$$

The required locus will be the eliminant of m in (3) and (4).

Putting the value m = (2a/k) from (4) in (3), we get

$$h = \frac{2a}{(2a/k)^2} \Rightarrow h = \frac{2ak^2}{4a^2} \text{ or } k^2 = 2ah$$

Generalising the required locus is $y^2 = 2ax$. **Proved.**

Example 28(b):

Find the equation of a common tangent of the parabola

$y^2 = 4ax$ and $x^2 = 4by$.

Solution:

The equation to the parabolas are given as

$$y^2 = 4ax \qquad ...(1)$$

and $$x^2 = 4by \qquad ...(2)$$

Equation to any tangent to (1) is $y = mx + a/m$. ...(3)

If this line is a tangent to (2) also, we must get only one point of intersection of (2) and (3).

Putting the value of y from (3) in (2), we get

$$x^2 = 4b\,(mx + a/m)$$

$$\Rightarrow \quad mx^2 - 4bm^2x - 4ab = 0. \qquad ...(4)$$

If (3) is a tangent to (2), (4) must be a perfect square.

So discriminant must be zero; hence

$$16b^2m^4 - 4m\,(-4ab) = 0$$

$$\Rightarrow \quad bm^4 + am = 0.$$

whence either m = 0

$$\Rightarrow \quad m = (-a/b)^{1/3}.$$

If m = 0, we do not get any finite value of a/m, so putting $m = (-1/b)^{1/3}$ in (3), the equation to the tangent becomes

$$y = (-1/b)^{1/3}\, x + a \div (-1/b)^{1/3}$$

$$b^{1/3}\, y + a^{1/3}\, x - a^{2/3}\, b^{2/3}$$ **Ans.**

Example 28(c):

Prove that two straight lines, one a tangent to the parabola $y^2 = 4a(x + a)$ and the other to the parabola $y^2 = 4a'(x + a')$, which are at right angles to one another, meet on the straight line $x + a + a' = 0$.

Show also that this straight line is the common chord of the two parabolas.

Solution:

The equations to the parabolas are given as

$$y^2 = 4a(x + a) \qquad \text{...(1)}$$

and $$y = 4a'(x + a') \qquad \text{...(2)}$$

Changing the origin to the $(-a, 0)$, (1) becomes

$$y^2 = 4ax. \qquad \text{...(3)}$$

Equation to any tangent to (3) is

$$y = mx + \frac{a}{m}.$$

Changing back to the first origin, this equation becomes

$$y = m(x + a) + \frac{a}{m}. \qquad \text{...(4)}$$

Example 29:

Find the focus and lotus rectum of the parabola

$$x^2 - 2ax + 2ay = 0.$$

Solution:

The equation is given as $x^2 - 2ax + 2ay = 0$

$$\Rightarrow (x - a)^2 = -2ay + a^2 = -4.a/2\,(y - a/2) \qquad \text{...(1)}$$

Changing the origin to (a.a/2) the equation (1) becomes

$$X^2 = 4(-a/2)Y \qquad \text{...(2)}$$

According to the new axis, the vertex of (2) is (0, 0), axis is Y-axis *i.e.* $X = 0$, latus rectum is $4a/2 = 2a$, and focus $(0, -a/2)$.

Changing back to the original axis, the co-ordinates of vertex will become (a,a/2), axis is $x - a = 0$ or $x = a$, the latus rectum is 2a and the focus is

$$(0 + a, -a/2 + a/2) \quad \text{or} \quad (a, 0)$$ **Ans.**

Example 30:

Find the latus rectum and focus of the parabola

$$y^2 = 4y - 4x.$$

Solution:

The given curve is $y^2 = 4y - 4x$

or $y^2 - 4y = -4x$ or $(y - 2)^2 = -4.\ 1.\ (x - 1)$ ...(1)

Changing the origin to (2, 1), equation No. (1) becomes

$$Y^2 = -4.\ 1X$$

where X and Y are new variables.

According to the new axes, the vertex to the parabola given by (2) is (0, 0), axis is axis of X, *i.e.* Y = 0, latus rectum is 4 and focus is (– 1, 0).

Changing back to the original axes the vertex will be (1, 2), axis is y – 2 = 0 or y = 2 latus rectum is 4 and the focus is (– 1 + 1, 0 + 2) or (0, 2). **Ans.**

Example 31:

For what point of the parabola $y^2 = 18x$ is the ordinate equal to three times the abscissa?

Solution:

Parabola is $y^2 = 18x$. ...(1)

Let (h, k) be any point on the parabola; then

$$k^2 = 18h. \quad ...(2)$$

If ordinate is 3 times the abscissa, k = 3h.

Putting the values in (2)

$$9h^2 = 18h \quad \text{or} \quad 9h^2 = 18h = 0 \quad \text{or} \quad 9h\ (h - 2) = 0.$$

Hence either h = 0 or h = 2 the corresponding values of k are 0 and 6.

Hence either h = 0 or h = 2 the corresponding values of k are 0 and 6.

Hence the required co-ordinates are (0, 0) or (2, 6). **Ans.**

Example 32:

In the parabola $y^2 = 6x$, find (1) the equation to the chord through the vertex and the negative end of the latus rectum, and (2) the equation to any chord through the point on the curve whose abscissa is 24.

Solution:

The given parabola is $y^2 = 6x$

or $y^2 = 4.\frac{3}{2}\ x.$

(i) Hence the value of a = 3/2; the focus will be (3/2, 0) and the vertex is clearly (0, 0). As the ends of the latus rectum are (a, ± 2a), the negative and will be (3/2, ± 3).

Equation of the chord joining origin (0, 0) and (3/2, – 3) is

$$y-0=\frac{-3-0}{\frac{3}{2}-3}(x-0) \quad \text{or} \quad 2x + y = 0. \qquad \textbf{Ans.}$$

(ii) Let the ordinate of the point whose abscissa is 24 be k so (24, k) will satisfy (1); so k^2 = 4.(3/2) 24 = 144 or k = 12, so point is (24, 12).

Equation of any line passing through (24, 12) is given as

$$y - 12 = m (x - 24). \qquad \textbf{Ans.}$$

Example 33:

Prove that the straight line x + y = 1 touches the parabola $y = x - x^2$.

Solution:

The equation of straight line is given as

$$x + y = 1. \qquad ...(1)$$

and the parabola as $y = x - x^2$. ...(2)

If (1) touches (2), then solving the two, we must get only one point of intersection. So putting the value of y from (1) in (2), we get

$$1 - x = x = x^2$$

or $x^2 - 2x + 1 = 0$

or $(x - 1)^2 = 0$ where x = 1.

As we get the two values of x coinciding, there is only one point of intersection.

Example 34:

If a chord of the parabola joining the points t_1, t_2 passes through the focus of the parabola, prove that $t_1t_2 = -1$.

Solution:

Let the equation of parabola be $y^2 = 4ax$. ...(i)

The equation of the chord joining t_1, t_2 is

$$(t_1 + t_2)\, y = 2x + 2at_1t_2. \qquad \text{(ii)}$$

Since it passes through the focus (a, 0) of parabola (i),

$\therefore \qquad 0 = 2a + 2a\, t_1 t_2$

or $\qquad t_1 t_2 = -1.$

Example 35(a):

A focal chord of the parabola $y^2 = 4ax$, meets it in point P and Q. If S be the focus, prove that

$$\frac{1}{SP} + \frac{1}{SQ} = \frac{1}{a}.$$

Solution:

The equation of parabola is $y^2 = 4ax$

Let $P(at^2, 2at)$ be a point and PSQ be the focal chord so that the co-ordinates of Q are $\left(\frac{a}{t^2}, \frac{-2a}{t}\right)$.

$\therefore$ SP = focal distance of point P = $a + x = a + at^2 = a(1 + t^2)$.

Similarly, $SQ = a + \frac{a}{t^2} = \frac{a(1+t^2)}{t^2}$.

$$\therefore \qquad \frac{1}{SP} = \frac{1}{SQ} = \frac{1}{a(1+t^2)} + \frac{t^2}{a(1+t^2)}$$

$$= \frac{(1+t^2)}{a(1+t^2)} = \frac{1}{a}.$$

Hence $\quad \frac{1}{SP} = \frac{1}{SQ} = \frac{1}{a}.$

Example 35(b):

Prove that the length of the intercept on the normal at the point $P(at^2, 2at)$ of the parabola $y^2 = 4ax$ made by the circle described on the line joining the focus and P as diameter is a $\sqrt{(1 = t^2)}$.

Solution:

The equation of the normal at $P(at^2, 2at)$ is given as

$$y + xt - 2at + at^3.$$

Let the circle with PS as diameter cuts the normal PG at B. This gives

$$PB = \sqrt{(PS^2 - SB^2)}$$

where $\quad PS = \sqrt{(at^2 - a)^2 + (2at - 0)^2}$

$$= \sqrt{a^2 (t^4 - 2t^2 + 1 + 4t^2)}$$

$$= a\ (t^2 + 1),$$

SB = Length of perpendicular from S(a, 0) on the line y + xt = 2at + at³

$$= \frac{-at + 2at + at^3}{\sqrt{(t^2 + 1)}}$$

$$= at\sqrt{(t^2 + 1)}.$$

Substituting the values of PS and SA in (i), we get

$$PB = a\sqrt{a^2 (t^2 + 1)^2 - a^2 t^2 (t^2 + 1)}$$

$$= a\sqrt{(t^2 + 1)}\left(\sqrt{t^2 + 1 - t^2}\right) = a\sqrt{t^2 + 1} \text{ (Hence the result follows).}$$

Example 36:

A parabola touches two given straight lines; if its axis passes through the point (h, k) the given lines being the axes of co-ordinates, prove that the locus of the focus is the curve $x^2 - y^2 - hx + ky = 0$.

Solution:

The focii are given by

$$ax = by = x^2 + y^2 + 2xy \cos w \qquad \text{...(1)}$$

and abscissa of the focus is

$$x = \frac{ab^2}{a^2 + 2ab \cos \omega + b^2}.$$

Since the axis goes through (h, k), we have

$$ak - bh = \frac{ab\left(a^2 - b^2\right)}{a^2 + 2ab \cos \omega + b^2} = \frac{x\left(a^2 - b^2\right)}{b}$$

Putting the values of a and b from (1), we get

$$\frac{k}{x} - \frac{h}{y} = xy\left(\frac{1}{x^2} - \frac{1}{y^2}\right); \text{ or } x^2 - y^2 + ky - bx = 0.$$

Example 37:

Prove that the straight line $y = mx + c$ touches the parabola $y^2 = 4a(x + a)$ if $c = ma + (a/m)$.

Solution:

The line is given as $y = mx + c$ ...(1)

and the parabola is $y^2 = 4a(x + a)$. ...(2)

To solve (1) and (2); we put the values of y from (1) in (2), so we get

$$(mx + c)^2 = 4a(x + a). \quad ...(3)$$

If (1) touches (2) the two values of m given by (3) must coincide, or it must be a perfect square, therefore its discriminant must be zero. Hence

$$\{2(mc - 2a)\}^2 - 4m^2(c^2 - 4a^2) = 0$$

$\Rightarrow$ $m^2c^2 + 4a^2 - 4amc - m^2c^2 + 4a^2m^2 = 0$

$\Rightarrow$ $4amc + 4a^2m^2 + 4a^2$

where $mc = am^2 + a$,

$\Rightarrow$ $c = am + a/m$. **Hence proved.**

Example 38:

A parabola touches two given straight lines at given points; prove that the locus of the middle point of the portion of any tangent which is intercepted between the given straight lines is a straight line.

Solution:

We know that $(x/f) + (y/g) = 1$ touches the parabola $\sqrt{(x/a)} + \sqrt{(y/b)} = 1$ if $(f/a) + (g/b) = 1$. (Ref. Article 242). The middle point will be $(f/2, g/2)$. Substituting the condition, the required locus is $(2x/a) + (2y/b) = 1$..

It is straight line. **Hence proved.**

Example 39:

TP and TQ are any two tangents to a parabola and the tangent at a third point R cuts them in P' and Q'; prove that

$$\frac{TP'}{TP} + \frac{TQ'}{TQ} = 1 \quad \textit{and} \quad \frac{QQ'}{QT'} = \frac{TP'}{P'P} = \frac{Q'R}{RP'}.$$

Solution:

We have know that, $(x/f) + (y/g) = 1$ ...(1)

always touches the curve $\sqrt{(x/a)} + \sqrt{(y/b)} = 1$ if $(f/a) + (g/b) = 1$...(2)

If TP, TQ be the axes and (1) the equation of the third tangent, then by (2),

$$\frac{TP'}{TP} + \frac{TQ'}{TQ} = 1$$ **Proved.**

From (2), $f/a = 1 - g/b = (b - g)/b$ ...(3)

and $g/b = 1 - f/a = (a - f)/a$ ...(4)

By (3) and (4), on division $\frac{b-g}{g} = \frac{f}{a-f}$.

As $\frac{QQ}{QT} = \frac{b-g}{g} = \frac{f}{a-f} = \frac{TP'}{PP'}$ if we draw RA parallel to TQ, then TH $= f^2/a$.

Hence $\frac{Q'R}{RP} = \frac{TH}{HP'} = \frac{f^2/a}{f - f^2/a} = \frac{f}{a-f} = \frac{TP'}{PP'}$. **Proved.**

Example 40:

From an external point P tangents are drawn to the parabola find the equation to the locus of P when these tangents make angles θ_1 and θ_2 with the axis, such that $\theta_1 + \theta_2$ is constant ($= 2\alpha$)

Solution:

$\theta_1 + \theta_2 = 2\alpha$, (by hypothesis); so $\tan(\theta_1 + \theta_2) = \tan 2\alpha$

$$\Rightarrow \quad \frac{\tan\theta_1 + \tan\theta_2}{1 - \tan\theta_1 \tan\theta_2} = \tan 2\alpha.$$

Putting the values from equations (2), we get

$$\frac{k/h}{1 - a/h} = \tan 2\alpha \text{ or } \frac{k}{h-a} = \tan 2\alpha.$$

Generalising and simplifying, we get the locus of (h, k) as

$$y = (x - a)\tan 2\alpha.$$

Example 41:

From an external point P tangents are drawn to the parabola find the equation to the locus of P when these tangents make angles θ_1 and θ_2 with the axis, such that $\tan^2\theta_1 + \tan^2\theta_2$ is constant ($= \lambda$).

Solution:

$$\tan^2\theta_1 + \tan^2\theta_2 = \lambda \text{ (by hypothesis)}$$

$$(\tan\theta_1 + \tan\theta_2)^2 - 2\tan\theta_1 \tan\theta_2 = \lambda.$$

Now we have

$$\frac{k^2}{h^2} - \frac{2a}{h} = \lambda \text{ or } k^2 - 2ah = \lambda h^2.$$

Generalising, we get the locus of (h, k) as $y^2 - 2ax = \lambda x^2$. **Ans.**

Example 42:

Two tangents to a parabola intercept on a fixed tangent segments whose product is constant; prove that the locus of their point of intersection is a straight line.

Solution:

Let $\left(at_1^2, 2at_1\right)$ and $\left(at_2^2, 2at_2\right)$ be the co-ordinates of the point of contact of the variable tangents respectively of the parabola.

$$y^2 = 4ax. \qquad ...(1)$$

Let $(at^2, 2at)$ be the fixed point A.

The co-ordinates of the point of intersection of variable tangents are given by $\quad x = at_1t_2 \qquad ...(2)$

and $\qquad y = a\,(t_1 + t_2). \qquad ...(3)$

Again, let B and C be $(att, a\,(t + t_1)$ and $(att_2, a\,(t + t_2)\}$ respectively as the co-ordinates of the points of intersection of fixed tangent with variable tangents.

Now, since BAC is a fixed tangent.

$\therefore$ BA $\times$ AC = constant say l (by hypothesis)

$$\Rightarrow \quad (at^2 - att_1)\,(at^2 - att_2) = l$$

$$\Rightarrow \quad a^2t^2\,(t - t_1)\,(t - t_2) = \lambda$$

$$\Rightarrow \quad (t - t_1)\,(t - t_2) = \lambda/a^2t^2$$

$$\Rightarrow \quad [t^2 - (t_1 + t_2)]\,t + t_1t_2] = \lambda/a^2t^2.$$

Substituting the values of $(t_1 + t_2)$ and t_1t_2, from (1) and (2), we get on generalisation

$$t^2 - \frac{y}{a} + \frac{x}{a} = \frac{\lambda}{a^2\,t^2} \quad \text{or } x - y = -\ \frac{\lambda}{a^2\,t^2} - at^2$$

which is a straight line. **Proved.**

Example 43(a):

Find the locus of the middle points of chords of the parabola which are normal to the curve.

Solution:

Let the chord PQ be normal at P ° $\left(at_1^2, 2at_1\right)$. If this intersects the parabola again at Q whose co-ordinates are $\left(at_2^2, 2at_2\right)$; then

$$t_2 = -t_1 - 2/t_1$$

or $$t_1 + t_2 = -\frac{2}{t_1} \qquad ...(1)$$

If (h, k) be the co-ordinates of the mid-point of PQ, then

$$h = \frac{1}{2}a\left(t_1^2 + t_2^2\right) \qquad ...(2)$$

and $$k\ (t_1 + t_2)\ a. \qquad ...(3)$$

To eliminate t_1 and t_2 between relations (1), (2) and (3), we have from (2) and (3) $-2/t_1$ = k/a or t_1 – 2a/k.

As $t_2 = -t_1 - 2/t_1$, we have $t_2 = \frac{k}{a} + \frac{2a}{k}$ or $t_2 = \frac{\left(k^2 + 2a^2\right)}{ak}$

Substituting for t_1 and t_2 in (2), we get $h = \frac{a}{2}\left[\frac{4a^2}{k^2} + \frac{\left(k^2 + 2a^2\right)^2}{a^2 k^2}\right]$.

$4a^4 + k^4 + 4a^4 + 4k^2a^2 = 2hk^2a$ or $k^2 (k^2 - 2ah + 4a^2) + 8a^4 = 0$.

Generalising the locus of (h, k) is $y^2 (y^2 - 2ax + 4a^2) + 8a^4 = 0$.

Proved.

Example 43(b):

Prove that the locus of the poles of tangents to the parabola $y^2 = 4ax$ with respect to the circle $x^2 + y^2 = 2ax$ is the circle $x^2 + y^2 = ax$.

Solution:

Let (h, k) be the co-ordinates of the pole; then polar of (h, k) with respect to the circle $x^2 + y^2 - 2ax = 0$ is

$$xh + yk - a\ (x + h) = 0$$

$$\Rightarrow \qquad y = x\left(-\frac{h}{k} + \frac{a}{k}\right) + \frac{ah}{k}. \qquad ...(1)$$

The equation to any tangent on the given parabola is

$$y = mx + \frac{a}{m}. \qquad ...(2)$$

Equations (1) and (2) represent the same straight line; hence equating the coefficients, we get

$$m = -\left(\frac{h}{k} + \frac{a}{k}\right) \quad m = \left(-\frac{h}{k} + \frac{a}{k}\right)$$

and $\frac{a}{m} = \frac{ah}{k}$.

To eliminate m between two equations, multiplying the two results, we get

$$\frac{ah}{k}\left(-\frac{h}{k}+\frac{a}{k}\right) = a \text{ or } h^2 + k^2 = ah.$$

Generalising, we get the required locus as $x^2 + y^2 = ax$. **Ans.**

Example 44:

A pair of tangents are drawn which are equally inclined to a straight line whose inclination to the axis is α; prove that the locus of their point of intersection is the straight line y (x – a) tan 2α.

Solution:

By hypothesis, the tangents through (h, k) are equality inclined with the line whose inclination with x-axis is α.

Then $\theta_2 - \alpha = \alpha - \theta_1$ or $\theta_1 + \theta_2 = 2\alpha$.

$\Rightarrow$ $\tan(\theta_1 + \theta_2) = \tan 2\alpha$

$$\Rightarrow \quad \frac{\tan\theta_1 + \tan\theta_2}{1 - \tan\theta_1 \tan\theta_2} = \tan 2\alpha = \frac{m_1 + m_2}{1 - m_1 m_2}$$

where m_1 and m_2 are the slopes of the tangents.

Putting the values of m_1 and m_2 we have

$$\frac{\left(\frac{k}{h}\right)}{\left(1-\frac{a}{h}\right)} = \tan 2\alpha \text{ or } \frac{k}{h-a} = \tan 2\alpha$$

Generalising and simplifying, the required locus of (h, k) is

$$y = (x - a) \tan 2\alpha.$$

Example 45:

From an external point P tangents are drawn to the parabola; find the equation to the locus of P when these tangents make angles θ_1 and q_2 with the axis, such that cos θ_1 cos θ_2 is constant (= μ).

Solution:

$$\cos\theta_1 \cos\theta_2 = \mu \text{ (by hypothesis)}$$

$$\Rightarrow \sec\theta_1 \sec\theta_2 = \frac{1}{\mu} \text{ or } \sec^2\theta_1 \sec^2\theta_2 = \frac{1}{\mu^2}$$

$$\Rightarrow \left(1+\tan^2\theta_1\right)\left(1+\tan^2\theta_2\right)=\frac{1}{\mu^2}.$$

$$\Rightarrow 1+\tan^2\theta_1+\tan^2\theta_2+\tan^2\theta_1\tan^2\theta_2=\frac{1}{\mu^2}$$

$$\Rightarrow 1+(\tan\theta_1+\tan\theta_2)^2-2\tan\theta_1\tan\theta_2+\tan\theta_2$$

$$+\tan^2\theta_1.\tan^2\theta_2=\frac{1}{\mu^2}.$$

From equations (2) and (3), Q.No. 1, we get $1+\frac{k^2}{h^2}-\frac{2a}{h}+\left(\frac{a}{h}\right)^2=\frac{1}{\mu^2}$

or $\quad h^2=\mu^2\{(h-a)^2+k^2\}$

Simplifying and generalising, the locus of (h, k) is

$$x^2=\mu^2\{(x-a)^2+y^2\}.$$

Example 46:

From an external point P tangents are drawn to the parabola; find the equation to the locus of P when these tangents make angles θ_1 and θ_2 with the axis, such than tan θ_1.tan θ_2 is constant (= c).

Solution:

We have tan θ_1.tan θ_2 = c (given).

We know that tan θ_1 tan θ_2 = a/h.

So a/h = c or hc = a. Generalising, the locus of (h, k) is cx = a. **Ans.**

Example 47:

From an external point P tangents are drawn to the parabola; find the equation to the locus of P when these tangents make angles θ_1 and θ_2 with the axis, such that cot θ_1 + cot θ_2 is constant (= d).

Solution:

cot q_1 + cot q_2 = d (by hypothesis)

$$\Rightarrow \quad \frac{1}{\tan\theta_1}+\frac{1}{\tan\theta_2}=d \text{ or } \frac{\tan\theta_2+\tan\theta_1}{\tan\theta_1\tan\theta_2}=d.$$

We get $\frac{k/h}{c/h}=d$ or k = ad.

Generalising, the locus of (h, k) is y = ad. **Ans.**

Example 48:

Prove that the locus of the point of intersection of two tangents which intercepts a given distance 4c on the tangent at the vertex is an equal parabola.

Solution:

Let (h, k) be the co-ordinates of the point of intersection of tangents on parabola $y^2 = 4ax$. ...(1)

The equation of any tangent is $y = mx + \frac{a}{m}$. ...(2)

We have $m_1 + m_2 = \frac{k}{h}$ ...(3)

and $m_1 m_2 = \frac{a}{h}$. ...(4)

Tangent at the vertex is y-axis *i.e.*, $x = 0$ ...(5)

Solving (2) and (5), points of intersection of tangents on y-axis are

$$\left(0, \frac{a}{m_1}\right) \text{ and } \left(0, \frac{a}{m_2}\right).$$

Now distance between the points $\left(0, \frac{a}{m_1}\right)$ and $\left(0, \frac{a}{m_2}\right)$.

$$= \frac{a}{m_2} - \frac{a}{m_1} = \frac{a(m_1 - m_2)}{m_1 m_2} = 4c \text{ (by hypothesis)}$$

$$\Rightarrow \quad \frac{a}{m_1 m_2} \sqrt{\left[(m_1 + m_2)^2 - 4m_1 m_2\right]} = 4c.$$

Putting the value from (3) and (4), we get

$$\frac{a}{a/h} \times \sqrt{\left(\frac{k^2}{h^2} - \frac{4a}{h}\right)} = 4c$$

$$\Rightarrow \quad h\sqrt{\left(\frac{k^2 - 4ah}{h^2}\right)} = 4c.$$

Squaring and simplifying, we get $k^2 - 4ah = 16c^2$.

Generalising, the locus of (h, k) is $y^2 - 4ax = 16c^2$

$$\Rightarrow \quad y^2 = 4a\,x + \left[\frac{4c^2}{a}\right].$$

This also a parabola of latus rectum 4a which is equal to the latus rectum of (1). Hence the two parabolas are equal. **Proved.**

Example 49:

If the normals at P and Q meet on the parabola, prove that the point of intersection of the tangents at P and Q lies either in a certain straight line, which is parallel to the tangents at the vertex, or on the curve whose equation is $y^2(x + 2a) + 4a^3 = 0$.

Solution:

Let the points P and Q be $\left(at_1^2, 2at_1\right)$ and $\left(at_2^2, 2at_2\right)$ and the equation of the parabola be y = 4ax. The equation of the tangents at P and Q are respectively $t_1y = x + at_1^2$ and $t_2y = x + at_2^2$.

Co-ordinates of point of intersection of these tangents are

$$\{at_1 t_2, a(t_1 + t_2)\}.$$

If the point of intersection is (h, k), then

$$h = at_1t_2 \quad ...(1)$$

and $$k = a(t_1 + t_2). \quad ...(2)$$

The equations to the normals at P and Q are respectively

$$y = -t_1x + 2at_1 + at_1^3$$

and $$y = -t_2x + 2at_2 + at_2^3.$$

As the point of intersection of these normals lies on the parabola, we have as in Question No. 28, Ex. 28, $t_1t_2 = 2$. ...(3)

From (1) and (3), x = 2a. It is a line parallel to x = 0, *i.e.*, y-axis.

Again. If the normals at P $\left(at_1^2, 2at_1\right)$ interest the parabola at Q $\left(at_2^2, 2at_2\right)$, then $t_2 = -t_1 - 2/t_1$

$$t_1 + t_2 = -2/t_1. \quad ...(4)$$

Now eliminate t_1 from the equations (1), (2) and (4), to find the locus of point of intersection of tangents. By (2) and (4), $k = -2a/t_1$ or $t_1 = -a/k$ and $t_2 = -hk/2a^2$ {from (1)}. Substituting for 't_1' and 't_2' in (2), we have

$$\frac{k}{a} = \frac{-2a}{k} - \frac{hk}{2a^2}.$$

Generalising and simplifying, we get $y^2(x + 2a) + 4a^3 = 0$.

Example 50:

Show the locus of the poles of tangents to the parabola $y^2 = 4ax$ with respect to the parabola $y^2 = 4bx$ is the parabola $y^2 = \frac{4b^2}{a}x$.

Solution:

Let (h, k) be the co-ordinates of the pole, then the polar of (h, k) with respect to parabola $y^2 = 4bx$ is

$$yk = 2b\,(x + h) \text{ or } y = \frac{2b}{k}x \times \frac{2bh}{k}. \qquad ...(1)$$

Equation to any tangent to $y^2 = 4ax$ is $y = mx + \frac{a}{m}$ *i.e.*, constant term is a/m.

So $$\frac{2bh}{k} = \frac{a}{2b/k} = \frac{ak}{2b} \text{ or } ak^2 = 4b^2\,h$$

Generalising, the locus of (h, k) is $y^2 = \frac{4b^2}{a}x$. **Proved.**

Example 51:

Show that the locus of the poles of chords which subtend a constant angle a at the vertex is the curve $(x + 4a)^2 = 4\cot^2\alpha\,(y^2 - 4ax)$.

Solution:

Let (h, k) be the co-ordinates of the polar of the variable chord.

Th equation of polar of (h, k) w.r.t. the parabola $y^2 = 4ax$ is

$$ky = 2a\,(x + h) \text{ or } \frac{xy - 2ak}{2ah} = 1. \qquad ...(1)$$

Now making the equation of parabola homogeneous with the help of (1), we get the equation of the straight lines joining the vertex (0, 0) to the points of intersection as

$$y^2 = 4ax\frac{ky - 2ax}{2ah} \text{ or } 4ax^2 - 2kxy + hy^2 = 0. \qquad ...(2)$$

If α is the angle between the lines given by (2), then

$$\tan\alpha = \frac{2\sqrt{(k^2 - 4ah)}}{4a + h} \quad \left(\text{as } \tan\theta = \frac{2\sqrt{(h^2 - ab)}}{a + b}\right)$$

$$\Rightarrow \qquad \tan^2\alpha = \frac{4\,(k^2 - 4ah)}{(4a + h)^2}$$

$\Rightarrow \qquad (h + 4a)^2 = 4 \cot^2 a\ (k^2 - 4ah).$

Generalising, the required locus is $(x + 4a)^2 = 4 \cot^2 a\ (y^2 - 4ax)$.

Example 52:

In the preceding question if the constant angle be a right angle the locus is a straight line perpendicular to the axis.

Solution:

Putting $\alpha = 90^\circ$ in last question, we get the locus of the pole as x =+ 4a = 0, $\therefore$ cot 90° = 0. This is a line perpendicular on x-axis, *i.e.*, y = 0.

Example 53:

A point P is such that the straight line drawn through it perpendicular to its polar with respect to the parabola $y^2 = 4ax$ touches the parabola $x^2 = 4by$. Prove that its locus is the straight line $2ax + by + 4a^2 = 0$.

Solution:

Let the equation to parabola be $y^2 = 4ax$ and the co-ordinates of the pole be (h, k); then the polar of the point (h, k) w.r.t. the parabola is yk = 2a (x + h). Then the equation of the line through (h, k) and perpendicular to its polar is

$$y - k = -\frac{k}{2a}(x - h)$$

$$\Rightarrow \qquad k\,(x - h) + 2a\,(y - k) = 0.$$

$$\Rightarrow \qquad x = \frac{-2a}{k}\,y + h + 2a \qquad \text{...(1)}$$

If (1) touches the parabola $x^2 = 3by$, it should be of the form

$$x = my + \frac{b}{m},$$

i.e., the constant term should be $\frac{b}{m}$.

$$\text{So} \qquad h + 2a = \frac{b}{-2a/k} = \frac{-kb}{2a}$$

$$\Rightarrow \qquad 2ah + kb + 4a^2 = 0$$

Generalising, the locus of (h, k) is $2ax + by + 4a^2 = 0$. **Proved.**

Example 54:

Find the locus of the middle points of chords of parabola which pass through the focus.

Solution:

Let the ends of chord of parabola $y^2 = 4ax$ be A and B whose co-ordinates are $\left(at_1^2, 2at_1\right)$ and $\left(at_2^2, 2at_2\right)$ respectively.

Equation of AB is $y(t_1 + t_2) = 2x + 2at_1t_2$. This passes through the focus (a, 0); so the co-ordinates will satisfy it, hence

$$2a + 2at_1t_2 = 0 \text{ or } t_1t_2 = -1. \quad ...(1)$$

Let (h, k) be the co-ordinates of the middle-point of the chord AB; then

$h = a\left(t_1^2 + t_2^2\right)/2$

or $\quad t_1^2 + t_2^2 = 2h/a \quad ...(2)$

and $\quad k = a(t_1 + t_2) \quad ...(3)$

The required locus will be obtained by eliminating t_1 and t_2 from (1), (2) and (3).

Identically, we have $(t_1 + t_2)^2 - \left(t_1^2 + t_2^2\right) = 2t_1t_2$

Putting the values from (1), (2) and (3), we get

$$\frac{k^2}{a^2} - \frac{2h}{a} = -2$$

$\Rightarrow \quad k^2 = 2a(h - a).$ **Ans.**

Example 55(a):

Through each point of the straight line $x = my + h$ is drawn the chord of the parabola $y^2 = 4ax$ which is bisected at the point; prove that it always touches the parabola $(y + 2am)^2 = 8a(x - h)$.

Solution:

Let the equation of the parabola be $y^2 = 4ax$. If (x_1, y_1) be the co-ordinates of the middle point of any chord, it will be parallel to the polar of (x_1, y_1) w.r.t. the parabola.

Equation of the polar of (x_1, y_1) w.r.t. the parabola is

$$yy_1 = 2a(x + x_1).$$

Its slope is $2a/y_1$; hence the slope of the chord, whose middle point is (x_1, y_1) also $2a/y_1$.

Hence the equation of the chord whose middle point is (x_1, y_1) is

$$\left(y - y_1\right) = \frac{2a}{y_1}\left(x - x_1\right)$$

$\Rightarrow \qquad y_1 (y - y_1) = 2a (x - x_1).$ ...(1)

Again as the point (x_1, y_1) lies on the line $x = my + h$ (given), so $x_1 = my_1 = h$.

Substituting the values of x_1 in (1), we have

$$y_1 (y - y_1) = 2a (x - my_1 - h)$$

$\Rightarrow \qquad (y + 2am) = \dfrac{2a}{y_1} (x - h) + \dfrac{2a}{2a / y_1}.$ ...(2)

Changing the origin to (–2am, h) and putting $2a/y_1 = M$, equation (2) becomes $Y = MX + 2a/M$ which is always tangent to the parabola

$Y^2 = 4.\ 2aX$ or $Y^2 = 8aX.$ ...(3)

As referred to the original (3) becomes

$(y + 2am)^2 = 8a (x - h).$ ...(4)

Hence (2) always touches (4). **Proved.**

Example 55(b):

Find the locus of the middle points of chords of the parabola which pass through the fixed point (h, k).

Solution:

Let $P \equiv (x_1, y_1)$ be the co-ordinates of the middle point of the chord AB of the parabola $y^2 = 4ax$.

Then we have, $\dfrac{2x_1}{a} = t_1^2 + t_2^2$ ...(1)

and $\qquad t_1 + t_2 = \dfrac{y_1}{a}.$ ...(2)

Equation of the chord AB is $y (t_1 + t_2) = 2x + 2at_1t_2$. As this passes through (h, k), co-ordinates will satisfy it; hence

$$k (t_1 + t_2) = 2h + 2at_1t_2 = 2h + a \{(t_1 + t_2)^2 - \left(t_1^2 + t_2^2\right)\}.$$

Putting the values from (1) and (2), we get

$$k.\frac{y_1}{a} = 2h + a \left\{\frac{y_1^2}{a^2} - \frac{2x_1}{a}\right\} = 3h + a \frac{\left(y_1^2 - 2ax_1\right)}{a^2}$$

$$\Rightarrow \qquad \frac{ky_1}{a} = \frac{2ah + \left(y_1^2 - 2ax_1\right)}{a}.$$

Generalising for (x_1, y_1), we get the locus of (x_1, y_1) as

$y^2 - ky = 2a (x - h).$ **Ans.**

Example 56:

Prove that the locus of the feet of the perpendiculars drawn from the vertex of the parabola upon chords, which subtend an angle of 45° at the vertex, is the curve $r^2 - 24ar\cos\theta + 16a^2\cos 2\theta = 0$.

Solution:

Let the equation to the parabola be given as

$$y^2 = 4ax \qquad ...(1)$$

and equation to any chord be, $x\cos\alpha + y\sin\alpha = p$. ...(2)

The equation to the lines joining the origin to the points of intersection of (1) and (2) is obtained by making (1) homogeneous with the help of (2).

So $$y^2 = \frac{4ax\left(x\cos\alpha + y\sin\alpha\right)}{p}$$

$\Rightarrow$ $$4a\cos\alpha\, x^2 + 4a\sin\alpha . xy - py^2 = 0.$$

Now as the angle between the lines is 45°, so

$$\tan 45^o = \frac{\pm\sqrt{\left(4a^2\sin^2\alpha + 4ap\cos\alpha\right)}}{4a\cos\alpha - p} \qquad ...(3)$$

as $$\tan\alpha = \frac{\pm 2\sqrt{\left(h^2 - ab\right)}}{a+b}.$$

Clearly the polar co-ordinates of the foot of the perpendicular from the vertex on chord (2) are (p, α).

Hence the required locus is obtained by generalising (3) for p and α.

So squaring, generalising and simplifying (3), the required locus is

$$16\,(a\sin^2\theta + r\cos\theta) = 16\,a^2\cos^2\theta + r^2 - 8\,are\cos\theta$$

or $$r^2 - 24ar\cos\theta + 16a^2\cos 2\theta = 0.$$ **Proved.**

Example 57:

Find the locus of the middle points of chords of the parabola which are of given length l.

Solution:

Let B and C be any two points $\left(at_1^2, 2at_1\right)$ and $\left(at_2^2, 2at_2\right)$ respectively on the parabola $y^2 = 4ax$; then

$$\text{Length BC} = \sqrt{\{(at_2^2 - at_1^2)^2 + (2at_2 - 2at_1)^2\}}$$
$$= a(t_2 - t_1)\sqrt{\{(t_1 + t_2)^2 + 4\}}$$

$\Rightarrow$ $BC = a\sqrt{\{(t_1 - t_2)^2 - 4t_1t_2\}}\ \sqrt{\{(t_1 - t_1)^2 + 4\}} = l$ ...(1)

Again, if P (h, k) be the mid-point of BC, then

$$\frac{2h}{a} = t_1^2 + t_2^2, \text{ and } t_1 + t_2\ \frac{k}{a} \text{ and as in last question,}$$

$$2t_1t_2 = \left\{(t_1 + t_2)^2 - \left(t_1^2 + t_2^2\right)\right\} = \frac{y_1^2}{a^2} - \frac{2x_1}{a} = \frac{y_1^2 - 2ax_1}{a^2}$$

Now by (1), $l^2 = a^2 \{(t_2 + t_1)^2 - 4t_1t_2\} \{(t_1 + t_2)^2 + 4]$

Putting the values of t_1 and t_2 etc, we get

$$l^2 = a^2 \left\{\frac{k^2}{a^2} - 2\frac{\left(k^2 - 2ah\right)}{a^2}\right\}\left(\frac{k^2}{a^2} + 4\right)$$

$$\Rightarrow \quad l^2 = \frac{\left(4ah - k^2\right)}{a^2}\left(k^2 + 4a^2\right).$$

Simplifying and generalising, we get the required locus as

$$(y^2 - 4ax)(y^2 + 4a^2) + a^2 l^2 = 0$$ **Ans.**

Example 58:

Two parabolas have the same axis and tangents are drawn to the second from points on the first; prove that the locus of the middle points of the chords of contact with the second parabola all lie on a fixed parabola.

Solution:

$$y^2 = 4x \quad ...(1)$$

and $$y = 4b(x + c) \quad ...(2)$$

be the equations of the parabolas and let PQ be the polar, the co-ordinates of ends P and Q to be $(at_1^2, 2at_1)$ and $(at_2^2, 2at_2)$ respectively; then the equation of polar PQ is $(t_1 + t_2)\,y = 2x + 2at_1t_2$ ...(3)

and if (x_1, y_1) be the pole, then polar of (x_1, y_1) w.r.t. (1) is

$$yy_1 = 2a(x + x_1) \quad ...(4)$$

Since (3) and (4) represent the same line, the coefficient will be proportional. So comparing the coefficients of x, y and constant term, we get

$$\frac{y_1}{t_1 + t_2} = \frac{2a}{2} = \frac{2ax_1}{2at_1t_2}$$

or $x_1 = at_1t_2$ and $y_1 = a(t_1 + t_2)$.

Hence the co-ordinates of the pole are $(at_1t_2, a(t_1 + t_2)\}$.

As this point lies on (2), we have

$$a^2(t_1 + t_2)^2 = 4b\,(at_1t_2 + c) = \frac{2b\left(2a^2t_1t_2 + 2ac\right)}{a}. \qquad ...(5)$$

If (h, k) be the co-ordinates of the middle point of the chord PQ, then

$$2h/a = t_1^2 + t_2^2 \text{ and } k/a = t_1 + t_2.$$

Hence $k^2 - 2ah = 2at_1t_2$

Substituting the values in (5)

$$a^2(k^2/a^2) = (2b/a)\,(k^2 - 2ah + 2ac).$$

Generalising, we have $ay^2 = 2b\,(y^2 - 2ax + 2ac)$ which is a fixed parabola.

Example 59:

Find the locus of the middle points of chords of the parabola which subtend a constant angle α at the vertex.

Section:

Let A(0, 0) be the vertex of the parabola $y^2 = 4ax$ and (h, k) be the co-ordinates of the middle point of the chord BC, where B and C are respectively $(at_1^2, 2at_1)$ and $(at_2^2, 2at_2)$; then

$$2h/a = t_1^2 + t_2^2 \qquad ...(1)$$

and $$k/a = t_1 + t_2 \qquad ...(2)$$

Now the slope of BA $= \dfrac{2at_1 = 0}{at_1^2 - 0} = \dfrac{2}{t_1} = m_1$ (say).

Similarly slope of CA $= \dfrac{2}{t_1} = m_1$ (say).

As the angle between AB and AC is given as a.

$$\pm \tan\alpha = \frac{m_2 - m_1}{1 + m_1m_2} = \frac{\dfrac{2}{t_2} - \dfrac{2}{t_1}}{1 + \dfrac{4}{t_1t_2}} = \frac{2(t_1 - t_2)}{t_1t_2 + 4}$$

$$= \frac{2\sqrt{\left\{(t_1 + t_2)^2 - 4t_1t_2\right\}}}{t_1t_2 + 4}. \qquad ...(3)$$

Putting the values of $(t_1 + t_2)$ and $2t_1t_2 = \{(t_1 + t_2)^2 - (t_1^2 + t_2^2)\}$, from (1) and (2), we get

$$\pm \tan\alpha = \frac{\sqrt{\left\{\dfrac{k^2}{a^2} - 2\left(\dfrac{k^2 - 2ah}{a^2}\right)\right\}}}{\left(\dfrac{k^2 - 2ah}{2a^2} + 4\right)}.$$

$$= \frac{2\sqrt{(4ah - k^2)} \times 2a}{(k^2 - 2ah + 8a^2)} = \frac{4a\sqrt{(4ah - k^2)}}{k^2 - 2ah + 8a^2}$$

$\Rightarrow$ $(k^2 - 2ah + 8a^2)^2 \tan^2 \alpha = 16a^2 (4ah - k^2)$ (on squaring).

Generalising, the required locus is

$$(8a^2 + y^2 - 2ax)^2 \tan^2 \alpha = 16a^2 (4ax - y^2).$$

Example 60:

Find the locus of a point O when the three normals drawn from it are such that two of them make equal angles with the given line $y = mx + c$.

Solution:

Let the line $y = mx + c$ make an angle θ with the axis; then $m = \tan \theta$. Let θ_1 and θ_2 be the angles of inclination of two normals having slopes m_1 and m_2 respectively; then

$$\theta_1 - \theta = \theta - \theta_2 \quad \text{[by hypothesis]}$$

$\therefore$ $\theta_1 + \theta_2 = 2\theta$ or $\tan(\theta_1 + \theta_2) = \tan 2\theta$

$$\Rightarrow \quad \frac{\tan\theta_1 + \tan\theta_2}{1 - \tan\theta_1 \tan\theta_2} = \frac{2\tan\theta}{1 - \tan^2\theta}$$

$$\frac{m_1 + m_2}{1 - m_1 m_2} = \frac{2m}{1 - m^2} = \lambda \text{ (say).}$$

Putting the values from equations (1) and (3)

$$m_1 + m_2 = -m_3 \text{ and } m_2 m_1 = \frac{-k}{am_3}.$$

$$\frac{-m_3}{1 + k/am_3} = \lambda \text{ or } \frac{-am_3^2}{k + am_3} = \lambda$$

$\Rightarrow$ $am_3^2 + alm_3 + k\lambda = 0.$ ...(1)

Again, $(m_1 + m_2)m_3 + m_1m_2 = \dfrac{2a - h}{a}.$

Similarly, we get the other two terms. Substituting in (1) we get

$$\frac{1}{am_3}\left(am_3^3 + 4k\right).\frac{1}{am_1}\left(am_1^3 + 4k\right).\frac{1}{am_2}\left(am_2^3 + 4k\right) = \lambda$$

$\Rightarrow$ $(4k + am_1^3)(4k + am_2^3)(4k + am_3^3) = \lambda a^3 m_1 m_2 m_3$

$\Rightarrow$ $64k^3 + 16k^2 a \Sigma m_1^3 + 4ka^2 \Sigma m_1^3 m_2^3 + a^2 m_1^3 m_2^3 m_3^3$

$= \lambda a^3 m_1 m_2 m_3.$ ...(2)

Now as $m_1 + m_2 + m_3 = 0$

$$\Rightarrow \quad m_1^3 + m_2^3 + m_3^3 = 3m_1m_2m_3 = -\frac{3k}{a}$$

$$\therefore \quad \Sigma m_1^3 = -\frac{3k}{a}.$$

Again $m_1m_2 + m_1m_3 + m_2m_3 = \frac{2a - h}{a}$;

$$\therefore \quad \{\Sigma(m_1m_2)\}^3 = \left(\frac{2a-h}{a}\right)^3$$

$$\text{L.H.S.} = \Sigma m_1^3m_2^3 + 3(m_1m_2 + (m_2m_3) \times (m_2m_3 + m_1m_2) \times (m_3m_1 + m_1m_2)$$

$$= \Sigma m_1^3m_2^3 + 3m_1m_2m_3 (m_1 + m_2)(m_2 + m_3)(m_3 + m_1)$$

$$= \Sigma m_1^3m_2^3 + 3m_1m_2m_3 -(m_3) -(m_1) -(m_2)$$

$$= \Sigma m_1^3m_2^3 - 3m_1^2m_2^2m_3^2.$$

Hence $\Sigma m_1^2m_2^2 = \frac{3k^2}{a^2} + \left(\frac{2a-h}{a}\right)^3$

Now putting different values in (2), we have

$$64k^3 + 16k^2a\left(-\frac{3k}{a}\right) + 4ka^2\left[\left(\frac{2a-h}{a}\right)^2 + \frac{3k^2}{a^2}\right] + a^3\left(-\frac{k^3}{a^3}\right) = \lambda a^3\left(-\frac{k}{a}\right)$$

or $\quad 64k^2 - 48k^3 + \frac{4k}{a}\,[(2a - h)^3 + 3ak^2] - k^3 + \lambda a^2k = 0$

$$\Rightarrow \quad 15k^3 + \frac{4k}{a}\,[(2a - h)^3 + 3ak^2] - k^3 + \lambda a^2k = 0$$

$$\Rightarrow \quad 15k^3 + \frac{4k}{a}\,(2a - h)^2 + 12k^3 + \lambda a^2k = 0$$

$$\Rightarrow \quad 27k^2a + 4(2a - h)^3 + \lambda a^3 = 0.$$

Generalising, the locus of (h, k) is

$$27ay^2 + 4(2a - x)^3 + \lambda a^3 = 0.$$ **Ans.**

$$\Rightarrow \quad -m_3^2 - \frac{k}{am_3} = \frac{2a-h}{a}$$

$$\therefore \quad am_3^2 + (2a - h)m_3 + k = 0. \qquad \ldots(2)$$

On subtracting (2) from (1) after multiplying (4) by m_3, we have

$$a\lambda m_3^2 + (k\lambda - 2a - h)\, m_3 - k = 0 \qquad \ldots(3)$$

Applying cross-multiplication to (2) and (3) to eliminate m_3, generalise the eliminant, which is the required to us and it is a parabola.

Example 61:

Through the vertex A of the parabola $y^2 = 4ax$ two chords AP and AQ are drawn and the circles on AP and AQ as diameters intersect in R. Prove that, if θ_1, θ_2 and ϕ be the angles made with the axis by the tangents at P and Q and by AR, then

$$\cot \theta_1 + \cot \theta_2 + 2 \tan \phi = 0.$$

Solution:

Let the equation of the parabola by $y^2 = 4ax$. Its vertex (say A) will be (0, 0).

Let P and Q be any two points on the parabola as

$$(at_1^2, 2at_1) \text{ and } (at_2^2, 2at_2)$$

Then the equation of the circle drawn on AP as diameter will be

$$x(x - at_1^2) + y(y - 2at_1) = 0$$

$$\Rightarrow \quad x^2 + y^2 - axt_1^2 - 2ayt_1 = 0 \qquad ...(1)$$

Similarly the equation of the circle having AQ as diameter is

$$x^2 + y^2 - axt_2^2 - 2ayt_2 = 0 \qquad ...(2)$$

(1) and (2) intersect at R and pass through A : so AR will be the common chord.

Equation of common chord AR will be

$$(x^2 + y^2 - axt_1^2 - 2ayt_1) - (x^2 + y^2 - axt_2^2 - 2ayt_2) = 0$$

$$\Rightarrow \quad x(t_1 + t_2) + 2y = 0$$

Hence the slope of AR $= -\dfrac{t_1 + t_2}{2} = \tan\phi$ (by hypothesis). ...(3)

Again tangent at P is $t_1 y = x + at_1^2$.

$\therefore$ Its slope $= 1/t_1 = \tan \theta_1$ (by hypothesis).

So $t_1 = \cot \theta_1$ ...(4)

Tangent at Q is $t_2 y = x + at_2^2$.

$\therefore$ Its slope $= 1/t_2 = \tan \theta_2$ (by hypothesis).

So $t_2 = \cot \theta_2$ ...(5)

By (3), (4) and (5)

$$2 \tan \phi = -(t_1 + t_2) = -\cot \theta_1 - \cot \theta_2$$

$$\Rightarrow \quad \cot \theta_1 + \cot \theta\theta_2 + 2 \tan \phi = 0.$$ **Proved.**

Example 62:

Prove that the orthocentres of the triangles formed by three tangents and the corresponding three normals to a parabola are equidistant from the axis.

Solution:

Let the parabola by $y^2 = 4ax$ ...(1)

and $(am_1^2, -2am_1)$, $(am_2^2, -2am2)$ and $(am_3^2, -2am_2)$ be the co-ordinates of the points P, Q and R respectively. Then the equations of the tangents at P, Q and R w.r.t. the parabola (1) are

$$-m_1y = x + am_1^2, \quad ...(2)$$

$$-m_2y = x + am_2^2, \quad ...(3)$$

$$-m_3y = x + am_2^2, \quad ...(4)$$

Solving (2) and (3), we have the co-ordinates of the point of intersection say A as $\{am_1m_2, -a(m_1 + m_2)\}$.

Now equations to the line through A and perpendicular to (3) will be

$$y + a(m_1 + m_2) = m_3(am_1m_3) \quad ...(5)$$

Similarly the equation to the line through B the point of intersection (3) and (4) and perpendicular to (2) is

$$y + a(m_2 + m_3) = m_1(x - am_2m_3) \quad ...(6)$$

On solving (5) and (6), we get the ordinate of the orthocentre of the triangle formed by (2), (3) and (4) as

$$-a(m_2 + m_2 + m_3 + m_1m_2m_3)$$

Again equations to the normal at P, Q and R are respectively

$$y = m_1x - 2am_1 - am_1^3, \quad ...(7)$$

$$y = m_2x - 2am_2 - am_2^3, \quad ...(8)$$

and $$y = m_3x - 2am_3 - am_3^2, \quad ...(9)$$

Solving (8) and (7), we have the co-ordinates of one of the vertice of the triangle formed by these three normals as

$$\{2a + a(m_1^2 + m_2^2 + m_1m_2), am_1m_2(m_1 + m_2)\}$$

Equation to the line through this point and perpendicular to (9) is

$$y - am_1 m_2(m_1 + m_2) = -\frac{1}{m_2}\left(x - 2a - am_1^2 - am_2^2 - am_1 m_2\right) \quad ...(10)$$

Similarly the equation to the other perpendicular will be

$$y - am_2 m_3(m_2 + m_3) = -\frac{1}{m_1}\left(x - 2a - am_2^2 - am_3^2 - am_2 m_3\right) \quad ...(11)$$

On solving (10) and (11), we get the ordinate of the point of intersection as $-a(m_1 + m_2 + m_3 + m_1m_2m_3)$.

This is also the ordinate of the point of intersection or ortho-centre of the triangle formed by tangents.

Hence the ortho-centres are equidistant from the axis i.e. x-axis.

Example 63:

If perpendiculars be drawn on any tangent to a parabola from two fixed points on the axis, which are equidistant from the focus, prove that the difference of their squares is constant.

Solution:

Let the equation to the parabola be $y^2 = 4ax$. ...(1)

Its axis is x-axis and focus is (a, 0). So any point on the axis equidistant from the focus may be taken as (a + k, 0) and (a – k, 0) say P and Q respectively.

Equation to any tangent to (1) is

$$y = mx + \frac{a}{m}. \qquad ...(2)$$

Distance of P from (2), say $PN = \dfrac{m(a+k)+a/m}{\sqrt{(1+m^2)}}$

and the distance of Q from (2) say $QM = \dfrac{m(a-k)+a/m}{\sqrt{(1+m^2)}}$

Hence $PN^2 - QM^2 = \dfrac{[m(a+k)+a/m]^2}{1+m^2} - \dfrac{[m(a-k)+a/m]^2}{1+m^2}$

$$= \frac{1}{(1+m^2)} m^2 (a+k)^2 + \frac{a^2}{m^2} + 2m.\frac{a}{m}(a+k) - m(a-k)^2 - \frac{a^2}{m^2} - 2\frac{a}{m}.m(a-k)\Big]$$

$$\frac{1}{(1+m^2)}[4m^2 ak + 4ak)$$

= 4ak which is constant as k and a are constant. **Hence proved.**

Example 64:

Find the equation to that tangent to the parabola $y^2 = 7x$ which is parallel to the straight line $4y - x + 3 = 0$. Find also its point of contact.

Solution:

The parabola is given as $y^2 = 7x$. ...(1)

here $4a = 7$ or $a = 7/4$.

Equation to any tangent to (1) is

$$y = mx + \frac{a}{m} \qquad ...(2)$$

The given line is $4y - x + 3 = 0$. ...(3)

Its slope = 1/4

If (2) is parallel to (3), m = 1/4. Substituting in (2), the equation of tangent is

$$y = \frac{1}{2}x + \frac{7}{4} \div \frac{1}{4}$$

$$\Rightarrow \quad 4y = x + 28.$$ **Ans.**

The point of contact is given by $\left(\frac{a}{m^2}, \frac{2a}{m}\right)$. So putting values of a and m, we get the point of contact as

$$\left\{\frac{7}{4} \div \left(\frac{1}{4}\right)^2, \ 2 \times \frac{7}{4} \div \frac{1}{4}\right\} \quad \text{or} \quad (28, 14)$$ **Ans.**

Example 65(a):

A tangent to the parabola $y^2 = 4ax$ makes an angle of 60° with the axis; find its point of contact.

Solution:

The curve is $y^2 = 4ax$. ...(1)

the axis of (1) is x-axis. If a line makes an angle of 60° to the axis, its slope is tan 60o = $\sqrt{3}$.

Any tangent to (1) is $y = mx + \frac{a}{m}$.

Putting the values of m as $\sqrt{3}$, we get the equation to the tangent as

$$y = \sqrt{3x} + \frac{a}{\sqrt{3}}$$

and the point of contact will be

$$\left(\frac{a}{m^2}, \frac{2a}{m}\right), \text{i.e.} \left(\frac{a}{3}, \frac{a}{\sqrt{3}}\right) \text{as } m = \sqrt{3}.$$ **Ans.**

Example 65(b):

Find the equation of tangent and normal. At the ends of the latus rectum of the parabola $y^2 = 4a\ (a - a)$.

Solution:

The equation to the parabola is given as

$$y^2 = 4a(x - a) \qquad ...(1)$$

Changing the origin to (a, 0), the equation of the curve becomes

$$y^2 = 4ax. \qquad ...(2)$$

Co-ordinates of the ends of latus rectum are (a, 2a) and (a, – 2a).

The equation to the tangent at (a, 2a) is

$$y.2a = 2a(x + a) \quad \text{or} \quad y = x + a$$

Changing the axes to the original origin, the equation will become

$$y = (x - a) + a \quad \text{or} \quad y = x.$$ **Ans.**

Equation of the normal according the new origin is

$$y - 2a = -\frac{2a}{2a}(x - a) \quad \text{or} \quad y + x = 3a.$$

Equation to the normal according to the old origin is

$$y + (x - a) = 3a \quad \text{or} \quad y + x = 4a$$ **Ans.**

Again, equation to the tangent at (a, – 2a) according to the new origin is $y(-2a) = 2a(x + a)$ or $x + y + a = 0$.

Equation to the normal at (a, – 2a) according to the new origin is

$$y - (-2a) = -\frac{-2a}{-2a}(x - a)$$

or $$y = x - 3a.$$

So equation to the normal at (a, – 2a) according to the old origin is

$$y = (x - a) - 3a \quad \text{or} \quad x - y = 4a$$ **Ans.**

Example 65(c):

Find the equation to the tangents to the parabola $y^2 = 9x$ which goes through.

Solution:

The parabola is given as $y^2 = 9x$. ...(1)

Equation to any tangent to (1) is

$$y = mx + \frac{9}{4m}\left(\text{as } a = \frac{9}{4}\right). \qquad ...(2)$$

If the tangent (2) passes through (4, 10), it will satisfy (2); hence

$$10 = m \cdot 4 + \frac{9}{4m}$$

$$\Rightarrow \quad 16m^2 - 40m + 9 = 0$$

$\Rightarrow$ $(4m = 9)(4m - 1) = 0$, whence $m = \frac{9}{4}$ or $m = \frac{1}{4}$.

If $m = 9/4$ substituting in (2), the equation to the tangent is

$$y = \frac{1}{4}x + \frac{9}{4}.\frac{4}{9} \quad \text{or} \quad 4y = 9x + 4.$$ **Ans.**

If $m = 1/4$, substituting in (2), we get

$$y = \frac{1}{4}x + \frac{9}{4}.\frac{4}{1} \quad \text{or} \quad 4y = x + 36.$$ **Ans.**

Example 66(a):

Find the points of the parabola $y^2 = 4ax$ at which (i) the tangent and (ii) the normal is inclined at 30° to the axis.

Solution:

The parabola is given as $y^2 = 4ax$. ...(1)

Any tangent to (1) is $y = mx + (a/m)$ and the corresponding point of contact is $\left(\frac{a}{m^2}, \frac{2a}{m}\right)$; as the axis of the parabola is x-axis, if the tangent is inclined to its axis at 30°, then $m = \tan 30^o = \frac{1}{\sqrt{3}}$. So the point of contact is

$$\left\{a / \left(\frac{1}{\sqrt{3}}\right)^2, 2a / \frac{1}{\sqrt{3}}\right\} \text{ or } \left(3a, 2\sqrt{3}\,a\right)$$ **Ans.**

Equation to any normal to (1) is $y = mx - 2am - am^3$ and the corresponding point of intersection is $(am^2, -2am)$.

It the normal is inclined at an angle of $30^o = 1/\sqrt{3}$.

So the corresponding point on the parabola is

$$\left\{a\left(\frac{1}{\sqrt{3}}\right)^2, -2a\,\frac{1}{\sqrt{3}}\right\}$$

$$\left(\frac{a}{3}, -\frac{2\sqrt{3}}{3}a\right)$$ **Ans.**

Example 66(b):

The normals at three points P, Q and R of the parabola $y^2 = 4ax$ meet in a point O whose coordinates are h and k; prove that the point O and the orthocentre of the triangle formed by the tangents at P, Q and R are equidistant from the axis.

Solution:

Let P, Q and R be respectively $\left(am_1^2,-2am_1\right)$, $\left(am_2^2,-2am_2\right)$ amd $\left(am_3^2,-2am_3\right)$.Then tangent at P is

$$-ym_1 = x + am_1^2 \qquad ...(1)$$

and tangent at Q is $-ym_2 = x + am_2^2$ ...(2)

Solving (1) and (2) simultaneously, the co-ordinates of the point of intersection are $x = am_1m_2$ and $y = -a(m_1 + m_2)$.

Tangent at R is $-ym_3 = x + am_3^2$. ...(3)

The equation to the line through [$amm_2 - a(m_1 + m_2)$] and perpendicular to (3) will be $y + a(m_1 + m_2) = m_3(x - am_1m_2)$

or $\qquad y = m_3x - am_1m_2m_3 - a(m_1 + m_2)$

or $\qquad y = m_3x + k + am_3$ ...(4)

Similarly the equation of the other perpendicular will be

$$y = m_2x + k + am_2. \qquad ...(5)$$

Solving (4) and (5), we get the co-ordinates of point of intersection *i.e.*, ortho-centre as (–a, k). As the ordinate of the ortho-centre and the point (h, k) are same, hence they are equi-distance from the x-axis.

Example 67:

Find the locus of a point O when the three normals drawn from it are such that two of them make angles with the axis the product of whose tangents is 2.

Solution:

$$m_1m_2 = 2 \qquad \text{[By hypothesis]...(1)}$$

$$m_1\,m_2\,m_3 = \frac{-k}{a}. \qquad ...(2)$$

By (1) and (2), $m_3 = \dfrac{-k}{2a}$. ...(3)

Again, we have

$$m_1\,m_2 - m_3^2 = \frac{2a-h}{a}.$$

Putting the values from (1) and (3), we have $2-\dfrac{k^2}{4a^2}=\dfrac{2a-h}{a}$,

whence $k^2 = 4ah$.

Generalising, the locus of (h, k) is $y^2 = 4ax$.

Example 68:

Find the locus of a point O when the three normals drawn from it are such that the area of the triangle formed by their feet is constant.

Solution:

Suppose the co-ordinates of the feet of the normals A, B and C be $\left(am_1^2 + 2am_1\right)$, $\left(am_2^2, -2am_2\right)$ and $\left(am_3^2, +2am_3\right)$ respectively.

Hence the area of the triangle ABC is given as

$$= \frac{1}{2}\left[am_1^2\left(-2am_2 + 2am_3\right) + am_2^2\left(-2am_3 + 2am_1\right) + am_3^2\left(-2am_1 + 2am_2\right)\right]$$

$$-a^2\left[m_1^2\left(m_2 - m_3\right) + m_2^2\left(m_3 - m_1\right) + m_3^2\left(m_1 - m_2\right)\right]$$

$\Rightarrow \quad a^2 (m_1 - m_2)(m_2 - m_3)(m_3 - m_1) = \text{constant (by hypothesis)}$

$\therefore \quad (m_1 - m_2)^2 (m_2 - m_3)^2 (m_3 - m_1)^2 = \text{constant} = \lambda \text{ (say)} \quad ...(1)$

But $(m_1 - m_2)^2 = (m_1 + m_2)^2 - 4m_1 m_2 = (-m_3)^2 + \dfrac{4k}{am_3}$

$$= \frac{1}{am_3}\left(am_3^3 + 4k\right)$$

Example 69:

The circle circumscribing the triangle PQR goes through vertex and its equation is $2x^2 + 2y^2 - 2x(h + 2a) - ky = 0$.

Solution:

Let the co-ordinates of P, Q and R are $\left(am_1^2, -2a\right)$ $\left(am_2^2, -2am_2\right)$ and $\left(am_3^2, -2am_3\right)$ respectively and let the equation of circle passing through P, Q and R be $x^2 + y^2 + 2gx + 2fy + c = 0$.

Since the circle passes through P, Q and R, substituting the ordinates one by one, we have

$$a^2m_1^4 + 3a^2m_1^2 + 2agm_1^2 - 4afm_1 + c = 0$$

$\Rightarrow \quad am_1\left(am_1^3 + 4am_1 + 2gm_1 - 4f\right) + c = 0$

and as $\quad am_1^3 + (2a - h)\, m_1 + k = 0$

We know that

$am_1^3 = (h - 2a)\, m_1 - k$; putting in (1), we get

$am_1 \{(h - 2a)\, m_1 - k + 4am_1 + 2gm_1 - 4f\} + c = 0$

$\Rightarrow a\ \{(h + 2a + 2g)\ m_2^2\ - (k + 4f)\ m_1\} + c = 0.$

Similarly by the points Q and R, we get

$$a\ \{(h + 2a + 2g)\ m_2^2 - (k+4f)\ m_2 + c = 0$$

and $$a\ \{(h + 2a + 2g)\ m_3^2\ - (k + 4f)\ m_3\} + c = 0$$

Subtracting (3) from (2), we get

$$a\ \{(h + 2a + 2g)\ \left(m_1^2 - m_2^2\right)\ - (k + 4f)\ (m_1 - m_2) = 0$$

$$\Rightarrow \quad (h + 2a + 2g)\ (m_1 + m_2) - (k + 4f) = 0\ (\text{as } m_1 \neq m_2)$$

and similarly from (3) and (4), we get

$$(h + 2a + 2g)\ (m_3 + m_2) - (k + 4f) = 0$$

$$(h + 2a + 2g)\ (m_3 + m_1) - (k + 4f) = 0$$

Adding (4), (5) and (6),

$$2\ (h + 2a + 2g)\ (m_1 + m_2 + m_3) - 3\ (k + 4f) = 0$$

But as $(m_1 + m_2 + m_3) = 0$ (by Q. No. 1), so $k + 4f = 0,\ 2f = \frac{-k}{2}$.

Subtracting (6) from (5), we get

$$h + 2a + 2g = 0 \text{ as } m_1\ '\ m_3 \text{ or } 2g = -\ (h + 2a).$$

Putting these values in (2), we get c = 0.

$$\Rightarrow \quad x^2 + y^2 - x\ (h + 2a) - \frac{k}{y} = 0$$

$$\Rightarrow \quad 2x^2 + 2y^2 - 2x\ (h + 2a) - ky = 0.$$

This passes through as the co-ordinates (0, 0) satisfy it.

Example 70:

Find the locus of a point O when the three normals drawn from it are such that the sum of the three angles made by them with the axis is constant.

Solution:

Let θ_1, θ_2 and θ_3 be the angles of inclination of normals with axis of x; then $\theta_1 + \theta_2 + \theta_3$ = constant (by hypothesis) = λ (say).

So $$\tan\ (\theta_1 + \theta_2 + \theta_3) = \tan \lambda$$

$$\Rightarrow \quad \frac{\tan \theta_1 + \tan \theta_2 + \tan \theta_3 - \tan \theta_1\ \tan \theta_2\ \tan \theta_3}{1 - \left(\tan \theta_1\ \tan \theta_2 + \tan \theta_2\ \tan \theta_3 + \tan \theta_1\ \tan \theta_3\right)} = \tan \lambda$$

$$\Rightarrow \quad \frac{m_1 + m_2 + m_3 - m_1\ m_2\ m_3}{1 - \left(m_1\ m_2 + m_2\ m_3 + m_1\ m_3\right)} = \tan \lambda.$$

Putting the values from equations No. (1), (2) and (3), we get

$$\frac{0-(-k/a)}{1-\frac{2a-h}{a}} = \tan \lambda$$

or $\frac{a}{h-a}$ = tan l or k = (h – a) tan λ.

Generalising, the locus of (h, k) is y = (x – a) tan α.

Example 71:

Find the locus of a point O when the three normals drawn from it are such that one bisects the angle between the other two.

Solution:

Let θ_1, θ_2 and θ_3 be the angles which the three normals make with x-axis; we have $\tan\theta_1 = m_1$, $\tan\theta_2 = m_2$ and $\tan\theta_3 = m_3$. Now one normal bisects the angle between two, we must have $\theta_1 - \theta_2 = \theta_2 - \theta_3$;

$\Rightarrow$ $2\theta_2 = \theta_1 + \theta_3$

$\Rightarrow$ $\tan 2\theta_2 = \tan(\theta_1 + \theta_3)$

$$\Rightarrow \quad \frac{2\tan\theta_2}{1-\tan^2\theta_2} = \frac{\tan\theta_1 + \tan\theta_3}{1-\tan\theta_1 \tan\theta_3}$$

$$\Rightarrow \quad \frac{2m_2}{1-m_2^2} = \frac{m_1+m_3}{1-m_1 m_3} = \frac{-m_3}{1-m_1 m_3}$$

$$\Rightarrow \quad m_2^2 + 2m_1m_3 = 3. \qquad ...(1)$$

Again, we have

$$m_1m_2 + m_2m_3 + m_1m_3 = \frac{2a-h}{a}$$

$$\Rightarrow \quad (m_1 + m_3)\, m_2 + m_1\, m_3 = \frac{2a-h}{a},$$

and as $m_1 + m_2 + m_3 = 0$

we get $-m_2^2 + m_1\, m_3 - \frac{2a-h}{a}$. ...(2)

From (1) and (2), by addition,

$$3m_1\, m_3 = \frac{2a-h}{a} + 3 = \frac{5a-h}{a} \text{ or } m_1\, m_3 = \frac{5a-h}{3a}$$

Putting this value in (4), we get $m_2^2 = \frac{2h-a}{3a}$

$$m_1\, m_2\, m_3 = \frac{-k}{a} \text{ or } m_1^2 m_3^2 = \frac{k^2}{a^2}.$$

Putting the values, we have $\frac{(5a-h)^2}{9a^2}.\frac{(2h-a)}{3a} = \frac{k^2}{a^2}.$

Generalising and simplifying, the locus of (h, k) is

$$(x - 5a)^2 (2x - a) = 27ay^2.$$

Example 72(a):

Find the locus of the middle points of chords of the parabola which are such that the normals at their extremities meet on the parabola.

Solution:

Since the normals at $P \equiv (at_2^1, 2at_1)$ and $Q \equiv (at_2^2, 2at_2)$ meet on the parabola $y^2 = 4ax$, the necessary condition is $t_1t_2 = 2$ ...(1)

Now, if (h, k) be the mid-point of the chord PQ, then as in Q. No. 19, we have

$$\frac{2x_1}{a} = t_1^2 + t_2^2 \qquad ...(2)$$

and $$\frac{y_1}{a} = t_1 + t_2. \qquad ...(3)$$

Now, we have to eliminate 't_1' 't_2' from (1), (2) and (3).

Identically, we have $(t_1 + t_2)^2 - (t_1^2 + t_2^2) = 2t_1t_2$.

Putting the values from (1), (2) and (3),

$$\frac{k^2}{a^2} - \frac{2h}{a} = 2a \text{ or } k^2 - 2ah = 4a^2.$$

Generalising and rearranging the required locus is $y^2 = 2a (x + 2a)$.

Ans.

Example 72(b):

For what point of the parabola $y^2 = 4ax$ (1) the normal equal to twice the subtangent, (2) the normal equal to the difference between the subtangent and the subnormal

Solution:

The parabola is given as

$$y^2 = 4ax \qquad ...(1)$$

If P be any point $(at^2, 2at)$ on the parabola, PT be the tangent at P meeting axis of x at T, PG be the normal meeting axis of x at G and PN be the perpendicular from P on X-axis, then TN is the subtangent, NG is the sub-normal and PG is the normal.

Equation to PT is $y.2at = 2a(x + at^2)$

$$\Rightarrow \qquad ty = x + at^2. \qquad ...(2)$$

Solving with X axis, i.e., putting

$$y = 0, \text{ we get } x = -at^2.$$

So co-ordinates of T are $(-at^2, 0)$.

Co-ordinates of N are clearly $(at^2, 0)$

Equation to the normal PG is

$$y - 2at = -\frac{2at}{2a}\left(x - at^2\right)$$

$$\Rightarrow \qquad y + tx = 2at + at^3$$

Again solving with X-axis i.e.

$$y = 0, \text{ we get}$$

$$x = 2a + at^2.$$

So the co-ordinates of G are $(2a + at^2, 0)$.

$$\text{Normal } PG = [(2a + at^2 - at^2)^2 + (0 - 2at)^2]^{1/2}$$

$$= 2a\sqrt{(1 + t^2)}.$$

(i) The condition is given as

normal = 2 × sub-tangent

$$\Rightarrow \qquad PG = 2TN$$

$$\text{or} \qquad PG^2 = 4TN^2$$

$$\Rightarrow \qquad 4a^2(1 + t^2) = 4(2at^2)^2. \qquad [\because TN = at^2 + at^2 = 2at^2]$$

$$\Rightarrow \qquad 4t^4 - t^2 - 1 = 0,$$

$$\text{whence} \quad t^2 = \frac{1 \pm \sqrt{(16+1)}}{2.4} = \frac{1 \pm 17}{8}$$

$$\Rightarrow \qquad t = \frac{1}{4}\sqrt{\left\{2\sqrt{17} + 2\right\}}.$$

(omitting -ve sign, as if t^2 is-ve, t will be imaginary),

So the co-ordinates of P are

$$\left[\frac{\sqrt{17}+1}{8}a, \frac{a}{2}\sqrt{\left\{2\sqrt{17}+2\right\}}\right]$$ **Ans.**

(ii) The condition is PG = TN – GN

$\Rightarrow \quad 2a\sqrt{(1 + t^2)} = 2at^2 - 2a$

$[\because GN = OG - ON = 2a + at^2 - at^2 = 2a]$

$\Rightarrow \quad \sqrt{(1 + t^2)} = (t^2 - 1).$

Squaring, we get $\quad 1 + t^2 = t^4 - 2t^2 + 1,$

whence $\quad t^2 = 0 \quad$ or $\quad t^2 = 3.$

If $t^2 = 0$, the co-ordinates of P become (0, 0) which is the vertex.

If $t^2 = 3$, the co-ordinates of P become $(3a, \sqrt{3}a)$. **Ans.**

Example 72(c):

Two equal parabolas have the same vertex and their axes are at right angles; prove that the common tangent touches each at the end at the and of a latus rectum.

Solution:

Taking the axis of the first parabola as x-axis and the axis of the other parabola as y-axis, their equation can be written as

$$y^2 = 4ax \quad \text{...(1)}$$

and $$x^2 = 4ay. \quad \text{...(2)}$$

(As the parabolas are equal, their latera recta are equal, say 4s).

Equation to any tangent to (1) is $y = mx + \frac{a}{m}$. ...(3)

To solve (2) and (3), we put the value of y from (3) in (2); we get

$$x^2 = 4a\left(mx + \frac{a}{m}\right)$$

$\Rightarrow \quad mx^2 - 4am^2x - 4a^2 = 0.$ ...(4)

If (3) touches (2), then (4) must be a perfect square, i.e. the discriminant of (4) must be zero. Hence

$$(-4am^2)^2 - 4.m(-4a^2) = 0$$

whence either $\quad m = 0$

or $\quad m^3 = -1$

or $\quad m = -1.$

If m = 0, line is x-axis, which is not a tangent to (1). So putting m = – 1 in (3), the equation to the common tangent is

$$y = -x - a$$

or $y + x + a = 0$ ...(5)

To get the point of contact, putting the value of m in (4),

$$-x^2 - 4ax - 4a^2 = 0$$

or $(x + 2a)^2 = 0$

whence $x = -2a.$

Putting this value in (2), y = a. So (5) touches (2) at (– 2a, a) which is one end of the latus rectum.

Similarly solving (1) and (5), we can show that the point of contact is (a, – 2a) which is the end of the latus rectum. **Proved.**

Example 72(d):

PN is an ordinate of the parabola; a straight line is drawn parallel to the axis to bisect NP and meets the curve in Q; prove that NQ meets the tangent at the vertex in a point T such that AT = (2/3) NP.

Solution:

Let the equation of the parabola be

$$y^2 = 4ax. \quad ...(1)$$

Tangent at the vertex is y-axis

or $x = 0$ ...(2)

Let P be any point $(at^2, 2at)$ on the parabola. So PN = 2at and co-ordinates of N are $(at^2, 0)$.

If M is the mid-point of PN, then NM = (1/2) PN = at.

As QM is parallel to x-axis, its equation is y = at. ...(3)

To solve (2) with (1), putting the value of y from (2) in (1), we get

$$a^2t^2 = 4ax \quad \text{or} \quad x = \frac{1}{4}at^2$$

So the co-ordinates of Q are $\left(\frac{1}{4}at^2, at\right)$

Equation of NQ will be

$$y - 0 = \frac{at - 0}{\frac{1}{4}at^2 - at^2}\left(a - at^2\right) \quad \text{or} \quad y = -\frac{4}{3t}\left(x - at^2\right) \quad ...(4)$$

If NQ meets the tangent at the vertex i.e. x = 0 at T, then putting x = 0 in (3), we get $y=\frac{4at}{3}$.

So the co-ordinates of T are = $0, \frac{4at}{3}t$ or $AT=\frac{4a}{3}t$.

Clearly AT = $\frac{2}{3}.2at=\frac{2}{3}.PN$. **Hence proved.**

Example 73:

A parabola is drawn touching the axis of x at the origin and having its vertex at a given distance k from this axis. Prove that the axis of the parabola is a tangent to the parabola $x^2 = -8k(y-2k)$.

Solution:

Let the equation to any parabola in its most general form be

$$(ax + by)^2 + 2gx + 2fy + c = 0. \quad ...(1)$$

As the parabola passes through the origin, hence c = 0, x-axis i.e. y = 0 is a tangent. So solving with (1), we get

$$a^2a^2 + 2gx = 0.$$

If y = 0 is a tangent, this equation must be a perfect square. So g = 0.

Putting the values of c and g, (1) becomes

$$(ax + by)^2 + 2fy = 0. \quad ...(2)$$

The axis of the parabola (2) will be

$$ax + by = -\frac{bf}{(a^2+b^2)} \quad ...(3)$$

To find out vertex, we solve (2) and (3) simultaneously.

Putting the value of ax + by from (3) in (2), we get

$$\frac{b^2f^2}{(a^2+b^2)}+2fy=0 \quad \text{or} \quad y=-\frac{b^2f}{2(a^2+b^2)^2}$$

So the ordinate of the vertex $-\frac{b^2f}{2(a^2+b^2)^2}=k$ (by hypothesis).

Hence $f=-\frac{2k(a^2+b^2)^2}{b^2}$

So (3) becomes

$$ax + by = -b\times\frac{-2k(a^2+b^2)^2}{b^2(a^2+b^2)}=\frac{2k(a^2+b^2)}{b}$$

$$\Rightarrow \qquad x=-\frac{b}{a}y+\frac{2k}{ab}\left(a^2+b^2\right)$$

$$\Rightarrow \qquad x=-\frac{b}{a}+2k\left(\frac{a}{b}+\frac{b}{a}\right) \qquad ...(4)$$

Equation to any tangent to the given parabola

$$x^2 = -8k(y-2k) \quad \text{or} \quad x^2 = 4.(-2k)(y-2k)$$

will be $x = m(y-2k) - \frac{2k}{m}$ $\qquad (\because a=-2k)$

$$\Rightarrow \qquad x = my - 2km - \frac{2k}{m}$$

$$\Rightarrow \qquad x = my - 2k\left(m+\frac{1}{m}\right) \qquad ...(5)$$

As it is true to any value of m, putting m = – b/a, (5) becomes

$$x = - \quad x=-\frac{b}{a}y-2k\left(-\frac{b}{a}-\frac{a}{b}\right)$$

$$\Rightarrow \qquad x=-\frac{b}{a}y+2k\left(\frac{b}{a}+\frac{a}{b}\right)$$

which is same as (4), the axis of the parabola. **Hence proved.**

Example 74:

PNP' is double ordinate of the parabola; prove that the locus of the point of intersection of the normal at P and the straight line through P' parallel to the axis is the equal parabola $y^2 = 4a(a - 4a)$.

Solution:

Let the equation to the parabola be $y^2 = 4ax$. ...(1)

If PQ be any double ordinate, let the co-ordinates of P be $(am^2, -2am)$; then the co-ordinates of Q will be $(am^2, 2am)$.

Equation to the normal at P is

$$y = mx - 2am - am^3. \qquad ...(2)$$

A line parallel to axis of parabola, i.e. x-axis through $Q \equiv (am^2, 2am)$ is

$$y = 2am \qquad ...(3)$$

The required locus will be obtained by eliminating m from (2) and (3). So putting the value of m from (3) in (2), we get

$$y=\frac{y}{2a}-2a\frac{y}{a}-a\left(\frac{y}{2a}\right)^3$$

Multiplying throughout by $8a^2$ and dividing by y, we get

$$8a^2 = 4ax - 8a^2 - y^2$$

$$\Rightarrow \quad y^2 = 4ax - 16a^2$$

$$\Rightarrow \quad y^2 = 4a(x - 4a) \text{ which is a parabola.}$$

Example 75:

Tangents are drawn to a parabola at points whose abscissa are in the ratio μ ; 1; prove that they intersect on the curve

$$y^2 = (\mu^{-1/4} + \mu^{-1/4})^2 ax.$$

Solution:

Let the equation to the parabola be $y^2 = 4ax$. ...(1)

Let P and Q be two points on the parabola (1), such that abscissa of P is μat^2 and of Q is at^2; then the corresponding ordinates are $2at\ \sqrt{\mu})$ and 2at. So the co-ordinats of P and Q are respectively $(\mu at^2, 2at\ \sqrt{\mu})$ and $(at^2, 2at)$.

Equation of tangent to (1) at $(\mu at^2, 2at\sqrt{\mu})$

$$y.2at\sqrt{\mu} = 2a\ (x + \mu at^2)$$

or $\quad ty\sqrt{\mu} = (x + \mu at^2)$. ...(2)

Similarly the equation to the tangent at Q is

$$ty = x + at^2.$$ ...(3)

The required locus of the point of intersection will be obtained by eliminating t from (2) and (3).

Multiplying (3) by μ and subtracting (2), we get

$$ty\sqrt{\mu}\ (1 - \sqrt{\mu}) = x\ (1 - \mu)$$

or $\quad t = \dfrac{x}{y} \cdot \dfrac{(1+\sqrt{\mu})}{\sqrt{\mu}}$ $\quad \left[\because 1 - \mu = 1(1-\sqrt{\mu})(1+\sqrt{\mu}\right]$

Substituting the value of t in (3) we get

$$y.\frac{x}{y}.\frac{\left(1+\sqrt{\mu}\right)}{\sqrt{\mu}} = x + a.\frac{x^2}{y^2}.\frac{\left(1+\sqrt{\mu}\right)^2}{\sqrt{\mu}}$$

Multiplying by μy^2, we get

$$xy^2\sqrt{\mu} + xy^2\mu = xy^2\mu + ax^2\ (1 + \sqrt{\mu})^2$$

or $\quad y^2\sqrt{\mu} = ax.\ (1 + \mu + 2\sqrt{\mu})$

or $$y^2 = ax\left[\frac{1}{\sqrt{\mu}} + \sqrt{\mu} + 2\right]$$

or $$y^2 = [\mu^{1/4} + \mu^{-1/4}]^2\ ax$$

which is required locus. **Proved.**

Example 76:

Prove that the straight line $lx + my + n = 0$ touches the parabola $y^2 = 4ax$ if $ln = am^2$.

Solution:

The straight line is given as

$$lx + my + n = 0 \quad ...(1)$$

and the parabola is given as $y^2 = 4ax$. ...(2)

The solve (1) and (2), we put the value of x from (1) in (2). So we get

$$y^2 = 4a\left[-\frac{my+n}{l}\right]$$

$$\Rightarrow \quad ly^2 + 4amy + 4an = 0. \quad ...(3)$$

If (1) touches (2), the roots of (3) must coincide, i.e. (3) must be a perfect square; so its discriminant will be zero.

Hence $(4am)^2 - 4l.4an = 0$

$\Rightarrow \quad a^2 m^2 = aln \quad$ or $\quad ln = am^2$ **Proved.**

Example 77:

The circle $x^2 + y^2 = 4ax$ and the parabola $y^2 = 4ax$.

Solution:

The equation to the circle is $x^2 + y^2 = 4ax$, ...(1)

and to the parabola is $y^2 = 4ax$. ...(2)

Equation to any tangent to (2) is $y = mx + a/m$ ...(3)

If (3) is a tangent to (1) also, the length of perpendicular from the centre of (1) on (3) must be equal to the radius of (1). Centre of (1) is (2a, 0) and radius is 2a,

Hence, we must have

$$\frac{m.2a - 0 + a/m}{\sqrt{(1+m^2)}} = 2a$$

$\Rightarrow \qquad 2m^2 + 1 = 2m\ \sqrt{(1 + m^2)}$

Dividing by m^2, we get $2 + \frac{1}{m^2} = \frac{2}{m}\sqrt{\left(1+m^2\right)}$

Squaring both sides, $4 = \frac{1}{m^4} = \frac{4}{m^2} = \frac{4}{m^2} + 4$

$\Rightarrow \qquad \frac{1}{m^4} = 0 \quad$ or $\quad \frac{1}{m} = 0.$

Dividing (3) m, we get $\frac{y}{m} = x + \frac{a}{m^2}.$

Putting the value of $1/m = 0$, we get the required equation as $x = 0$

Example 78(a):

From the proceeding question prove that, if tangents be drawn to the parabola $y^2 = 4ax$ from any point on the parabola $y^2 = a(x + b)$, then the normals at the points of contact meet on a fixed straight line.

Solution:

Let (x', y') and (x", y") be any two points on the parabola $y^2 = 4ax$ and let the tangents on (x', y') & (x", y") meet in (x_1, y_1), then by hypotesis, (x_1, y_1) lies on the parabola $y^2 = a(x + b)$. So the co-ordinates will satisfy it. Hence we get

$$y_1^2 = a(x_1 + b). \qquad ...(1)$$

Let the normals at (x', y') and (x", y") meet in (h, k); then by

$$h = 2a + \frac{y_1^2}{a} + x_1 \qquad ...(2)$$

and

$$k = -\frac{x_1 y_1}{a} \qquad ...(3)$$

(putting h and k for x_2 and y_2 respectively).

The required locus will be obtained by eliminating (x_1, y_1) from (1), (2) and (3).

Putting the values of y_1^2 from (1) in (2), we have

$$h = 2a + \ h = 2a + \frac{1}{a}a\left(x_1 + b\right) - x_1$$

$\Rightarrow \qquad h = 2a + x_1 + b - x_1 = 2a + b.$

Generalising, we get $x = 2a + b$ which is a fixed line. So the tangents meet on a fixed line.

Example 78(b):

Prove that the chord of the parabola $y^2 = 4ax$, whose equation is $y - x\sqrt{2} + 4a\sqrt{2} = 0$, is a normal to the curve and that its length is $6\sqrt{3}a$.

Solution:

The equation to the parabola is given as

$$y^2 = 4ax, \qquad ...(1)$$

and the equation to the chord is

$$y - x\sqrt{2} + 4a\sqrt{2} = 0. \qquad ...(2)$$

To solve (1) and (2) simultaneously, we put the value of y from (2) in (1); we get $(x\sqrt{2} - 4a\sqrt{2})^2 = 4ax$.

$$\Rightarrow \quad 2x^2 + 32a^2 - 16ax = 4ax$$

$$\Rightarrow \quad x^2 - 10ax + 16a^2 = 0.$$

whence either $x = 8a$

or $\quad x = 2a$.

Putting in (2), the corresponding values of y are respectively $4a\sqrt{2}$ and $-2a\sqrt{2}$. Hence the points of intersection of (1) and 92) are $(8a, 4a\sqrt{2})$ and $(2a, -2a\sqrt{2})$

The distance between the points $(8a - 4a\sqrt{2}$ and $(2a - 2a\sqrt{2})$

$$\sqrt{\{(8a - 2a)^2 + (4a\sqrt{2} + 2a\sqrt{2})^2\}}$$

$$\sqrt{\{(36a^2 + 72a^2} = 6a\sqrt{3}. \qquad \textbf{Ans.}$$

Hence the length of the chord $= 6a\sqrt{3}$.

Again, equation of the normal at $(8a, 4a\sqrt{2})$ is

$$y - 4a\sqrt{2} = \frac{4a\sqrt{2}}{2a}(x - 8a)$$

$$\Rightarrow \quad 2x\sqrt{2} - y = 12a\sqrt{2}$$

And the normal at $(2a, -2a\sqrt{2})$is

$$y + 2a\sqrt{2} = \frac{-2a\sqrt{2}}{2a}(x - 2a)$$

$$\Rightarrow \quad y - x\sqrt{2} + 4a\sqrt{2} = 0$$

Which is same as (2), hence the chord given by (2) is a normal at $(2a, -2a\sqrt{2})$ **Proved.**

Example 78(c):

It P, Q and R be three points on a parabola whose ordinates are in geometrical progression, prove that the tangents at P and R meet on the ordinate of Q.

Solution:

Let the equation of the parabolas $y^2 = 4ax$. ...(1)

and the co-ordinates of three points P, Q and R on it be $(at_1^2, 2at_1)$, $(at_2^2, 2at_2)$ and $(at_3^2, 2at_3)$ respectively. If the ordinates are in GP, then

$$2at_1^2 \times 2at_3 = (2at_2)^2 \quad \text{or} \quad t_1 t_3 = t_2^2$$

Equation to the tangent on (1) at $P \equiv (2at_1^2, 2at_1)$ is

$$y.2at_1 = 2a (x + at_1^2) \quad \text{or} \quad yt_1 = x + at_1^2. \quad ...(3)$$

Similarly the equation to the tangent at $R \equiv (2at_3^2, 2at_3)$ to (1) is

$$yt_3 = = x + at_3^2. \quad ...(4)$$

To get the value of x from (3) and (4), multiplying (3) by t_3, (4) by t_1 and subtracting, we get

$$0 = x (t_3 - t_1) - at_1t_3 (t_3 - t_1)$$

$$\Rightarrow \quad x = at_3t_1 = at_2^2 \quad [\text{by } (2), t_1t_3 = t_2^2]$$

The abscissa of Q is also given by $x = at_2^2$. **Hence proved.**

Example 78(d):

Prove that the distance between a tangent to the parabola and the parallel normal is a cosec θ sec^2 θ, where θ is the angle the either makes with the axis.

Solution:

Let any parabola be $y^2 = 4ax$. ...(1)

And tangent to (1) will be $y = mx + \frac{a}{m}$. ...(2)

If there be any normal which is parallel to (2), the slopes will be the same; so the equation of the normal parallel to (2) will be

$$y = mx - 2am - \theta m^3. \quad ...(3)$$

If ON' be the lengths of the perpendiculars from the origin O upto (2) and (3) respectively, clearly, the distance between (2) and (3) is ON – ON'. So the distance

$$ON - ON' = \frac{a/m}{\sqrt{(1+m^2)}} - \frac{-2am - am^2}{\sqrt{(1+m^2)}}$$

If θ is the angle which (1) or (2) make with x-axis, then

Putting the values, m = tan θ

$$ON - ON' = \frac{a/\tan\theta}{\tan\theta\sqrt{(1+\tan^2\theta)}} - \frac{-2\tan\theta - a\tan^2\theta}{\sqrt{(1+\tan^2\theta)}}.$$

$$= \frac{a}{\tan\theta\sec\theta}\left[1+2\tan^2\theta+\tan^4\theta\right]$$

$$= \frac{a\cos^2\theta}{\sin\theta}.\left[1+\tan^2\theta\right]^2$$

$= a$ cosec θ. $\cos^2\theta \sec^2\theta = a$ cosec θ $\sec^2\theta$. **Proved.**

Example 79:

Prove that two parabolas, having the same focus and their axes in opposite directions, cut at right angles.

Solution:

Let us take the common focus as the origin, x-axis as the common axis (in opposite direction), the line perpendicular to the axis at 0 as y-axis, and the latus rectum of the parabolas as 4a and 4b, then their equations may be written as

$$y^2 = 4a(x + a)$$

or $\quad y^2 = 4ax + 4a^2$ ...(1)

and $\quad y^2 = -4b(x - b)$

or $\quad y^2 = -4bx + 4b^2$ ...(2)

To get the points of intersection, we solve (1) and (2). So subtracting (2) from (1),

$$0 = 4(a + b)x + 4(a^2 - b^2)$$

or $\quad x = (b - a)$.

Putting this value in (1), $y^2 = 4b(b - a) + 4a^2 = 4ab$

$\Rightarrow \quad y = \pm 2\sqrt{(ab)}$.

Hence the points of intersection (say P and Q) are $\{b - a, 2\sqrt{(ab)}\}$ and $\{b - a, 2\sqrt{(ab)}\}$.

Tangent at P to (1) will be

$y.2\sqrt{(ab)} = 2a(x + b - a) + 4a^2$. ...(3)

Tangent at P to (2) will be

$y.2\sqrt{(ab)} = -2b(x + b - a) + 4b^2$. ...(4)

Slope of (3) is $\frac{2a}{2\sqrt{(ab)}} = \sqrt{\frac{a}{b}} = m_1$ (say)

and slope of (4) is $-\frac{2b}{2\sqrt{(ab)}} = \sqrt{\frac{b}{a}} = m_2$ (say)

Then $m_1 \times m_2 = \sqrt{\frac{a}{b}} \times -\sqrt{\frac{b}{a}} = -1$. So the two tangents are perpendicular to each other.

By symmetry, the tangents at Q are also perpendicular to each other.

Hence the two curves (1) and (2) intersect at right angles.

Example 80:

Find the lengths of the normals drawn from the point on the axis of the parabola $y^2 = 8ax$ whose distance from the focus is 8a.

Solution:

The parabola is $y^2 = 8ax = 4.\ 2ax$. ...(1)

Hence the focus is (2a, 0). Clearly the co-ordinates of the point which is at a distance of 8a from (2a, 0) and is on the axis of the parabola i.e. x-axis will be (10a, 0).

Equation to any normal to the given parabola is

$$y = mx - 2\ .\ 2a.m - 2am^3. \quad ...(3)$$

If the normal passes through (10a, 0), the co-ordinates will satisfy. So $0 = 10am - 4am - 2am^3$.

or $2am\,(3 - m^3) = 0$

whence $m = 0$

or $m = \pm \sqrt{3}$.

If m = 0, it is x-axis; so length of the normal is 10a clearly.

The point of intersection of (2) and (1) is $(2am^2, -4am)$.

Putting the value of $m = \sqrt{3}$, the point is $(6a, -4a\sqrt{3})$.

So the length of the normal

$$= \sqrt{\{(10a - 6a)^2 + (0 + 4a\sqrt{3})^2} = \sqrt{(16a^2 + 48a^2)} = 8a.$$

Similarly, putting $m = -\sqrt{3}$ length will be 8a.

Example 81:

Prove that the locus of the middle point of the portion of a normal intersected between the curve and the axis is a parabola whose vertex is the focus and whose latus rectum is one quarter of that of the original parabola.

Solution:

Let $y^2 = 4ax$ be any parabola and point P on it be $(am^2, -2am)$.

Then the equation to the normal at $(am^2, -2am)$ for the given parabola will be $y = mx - 2am - am^3$. ...(1)

If (1) meets the axis of parabola i.e x-axis at $(x_1, 0)$, then these co-ordinates will satisfy (1). So $0 = mx_1 - 2am - am^2$

or $x_1 = (2a = am^2)$.

Hence the point of intersection say G becomes

$(2a + am^2, 0)$.

Let (h, k) be mid-point of PG; then

$$h = \frac{am^2 + (2a + am^2)}{2} = a + am^2 \quad ...(2)$$

and $$k = \frac{0 - 2am}{2} - am \quad ...(3)$$

The required locus will be obtained by eliminating m from (2) and (3).

So putting the values of m from (3) in (2), we get

$$h = a + a\left(-\frac{k}{a}\right)^2$$

Generalising and simplifying, we get the required locus as

$y^2 = a(x - a)$

This is a parabola whose vertex is (a, 0), i.e. focus of (1) and latus rectum is a, i.e. 1/4 of that of (1). **Hence proved.**

Example 82:

Show that the two parabolas

$x^2 + 4a(y - 2b - a) = 0$ *and* $y^2 = 4b(x - 2a + b)$

intersect at right angles at the common end of the latus rectum of each.

Solution:

The two parabolas are given as

$x^2 + 4a(y - 2b - a) = 0$

or $x^2 + - 4a[y - (a + 2b)]$ ...(1)

and $y^2 = 4b(x - 2a + b)$

or $y^2 = 4b[x - (2a - b)]$ ...(2)

For (1) changing the origin to {0, (a + 2b)}, we find the ends of the latus rectum, say P and Q as (2a, a) and (– 2a, a). Hence with respect to the original origin, the co-ordinates of P are [2a, a + 2b – a] or (2a, 2b).

Similarly the co-ordinates of one of the end points of the latus rectum in (2) is (2a, 2b).

Hence (2a, 2b) is the common point of (1) and (2) which is one of the end points of the latus rectum.

Now the equations to the tangent at (2a, 2b) to (1) and (2) are respectively

$$x.2a = -2a\,[y - 2b - (a + 2b)] \quad ...(3)$$

$$x.2b = 2b\,[x - 2a - (2a + b)]. \quad ...(4)$$

The slope of (3) and (4) are respectively – 1 and 1.

Hence their product is – 1, or the tangents are perpendicular to each other.

So the two curves intersect at right angles

Example 83:

The normal at any point P meets the axis in G and the tangent at the vertex in G'; if A be the vertex and the rectangle AGQG' be completed, prove that the equation to the locus of Q is

Solution:

Let P be any point $(at^2, 2at)$ on the parabola $y^2 = 4ax$. ...(1)

Equation of the normal PG at $(at^2, 2at)$ tp (1) is

$$y = -tx + at + at^3 \quad ...(2)$$

[as the normal at $(am^2, -2am)$ is

$y = [mx - 2am - am^3]$.

If G is the point on x-axis, ordinate of G will be 0; so its abscissa will be $(2a + at^2)$.

Here $AG = 2a + at^2$. ...(3)

Similarly if PG meets tangent at the vertex i.e. y-axis at G'; abscissa of G' is zero, so ordinate will be [by putting x = 0 in (2)].

$$y = 2at + at^3$$

So $AG = 2at + at^3$. ...(4)

If the required point Q be (h, k), clearly by the fig.,

$$h = AG = 2a + at^2 \text{ [by (3)]} \quad ...(5)$$

$$k = AG' = 2a + at^2 \text{ [by (3)]} \quad ...(6)$$

The required locus will be obtained by eliminating t from (5) and (6). So multiplying (5) by t and subtracting from (6); we get

$$k - ht = 0, \text{ so } t = \frac{k}{h}.$$

Putting in (5), we get $h = 2a + a\left(\frac{k}{h}\right)^2$

$\Rightarrow \quad h^3 = 2ah^2 + ak^2$

Generalising we get $x^3 = 2ax^2 + ay^2$. **Proved.**

Example 84:

Prove that the parabolas $y^2 = ax$ and $x^2 = by$ cut one another at an angle $\tan^{-1}\frac{3a^{1/3}b^{1/3}}{2\left(a^{2/3}+b^{2/3}\right)}$

Solution:

The parabolas are given as

$$y^2 = ax \qquad ...(1)$$

and $$x^2 = by \qquad ...(2)$$

To solve (1) and (2), putting the value of y from (2) in (1), we get

$$\frac{x^4}{b^2} = ax.$$

whence either $x = 0$ or $x = a^{1/2}b^{2/3}$. The corresponding values of y are) and $a^{1/3}b^{1/3}$. Hence the points of intersection of (1) and (2) are (0, 0) and $(a^{1/3}b^{2/3}, a^{2/3}b^{1/3})$.

The tangents at (0, 0) to (2) and (1) are respectively the x and y axis, so the angle between them is 90°.

Now tangents at $(a^{1/3}b^{2/3}, a^{2/3}b^{1/3})$ to (1) and (2) are respectively

$$y.\ a^{2/3}\ b^{1/3} = \frac{a}{2}.\left(x + a^{1/3}.b^{2/3}\right) \qquad ...(3)$$

and $$y.\ a^{1/3}\ b^{2/3} = \frac{b}{2}.\left(y + a^{2/3}.b^{1/3}\right) \qquad ...(4)$$

Slopes of (3) and (4) are respectively

$$\tan\theta = \frac{2.\frac{a^{1/3}}{b^{1/3}} - \frac{1}{2}.\frac{a^{1/3}}{b^{1/3}}}{1 + 2.\frac{a^{1/3}}{b^{1/3}}.\frac{a^{1/3}}{2b^{1/3}}} = \frac{\frac{3}{2}\frac{a^{1/3}}{b^{1/3}}}{\frac{\left(b^{2/3}+a^{2/3}\right)}{b^{2/3}}}$$

$$= \frac{3a^{1/3}b^{1/3}}{2\left(b^{2/3}+a^{2/3}\right)}$$

or $$\theta = \tan^{-1}\left\{\frac{3a^{1/3}b^{1/3}}{2\left(a^{2/3}+b^{2/3}\right)}\right\}$$ **Proved.**

Example 85(a):

Prove that the two parabolas $y^2 = 4ax$ and $y^2 = 4ax\ (x - b)$ cannot have a common normal, other than the axis, unless $b/(a - c) > 2$.

Solution:

The two parabolas are given as

$$y^2 = 4ax \quad ...(1)$$

and $$y^2 = 4c\ (a - b). \quad ...(2)$$

Equation to any normal to (1) is

$$y = mx - 2am - am^3. \quad ...(3)$$

Equation to any normal to (2) is

$$y = m\ (x - a) - 2cm - am^3. \quad ...(4)$$

If there is any common normal, then (3) and (4) must be identical. As the coefficients of x and y are equal, so the constant terms will also be equal; hence

$$-2am - am^3 = -bm - 2mc - cm^3$$

$$\Rightarrow \quad m\ [m^2\ (c - a) - 2a + b + 2c)] = 0.$$

So either $m = 0$

or $$m^2 = \frac{2a-b-2c}{c-a}.$$

If $m = 0$, the common normal is the x-axis.

If $m^2 = \frac{2a-b-2c}{c-a}$, then

$$m = \sqrt{\left(\frac{2(a-c)-b}{c-a}\right)} = \sqrt{\left(-2-\frac{b}{c-a}\right)}$$

If the value of m is real, then $-2 - \frac{b}{c-a} > 0$

$$-\frac{b}{c-a} \geq 2 \quad \text{or} \quad \frac{b}{c-a} \geq 2.$$

Hence proved.

Example 85(b):

Prove that three tangents to a parabola, which are such that the tangents of their inclinations to the axis are in a given harmonical progression, form a triangle whose area is constant.

Solution:

Let the parabola be $y^2 = 4ax$ and say any three tangents on it are

$$y = m_1 x + \frac{a}{m_1} \qquad ...(1)$$

$$y = m_2 x + \frac{a}{m_2} \qquad ...(2)$$

and $$y = m_3 x + \frac{a}{m_3} \qquad ...(3)$$

Solving (1) and (2), we get

$$x = \frac{a}{m_1 m_2} \quad \text{and} \quad y = \frac{a}{m_1} + \frac{a}{m_2}$$

Hence the points of intersection of (1) and (2), (2) and (3) and (1) and (3) respectively, i.e., the co-ordinates of the angular points of the triangle are

$$\left(\frac{a}{m_1 m_2}, \frac{a}{m_1} + \frac{a}{m_2}\right), \left(\frac{a}{m_2 m_3}, \frac{a}{m_2} + \frac{a}{m_3}\right) \text{and} \left(\frac{a}{m_1 m_3}, \frac{a}{m_3} + \frac{a}{m_1}\right)$$

Hence the area of the triangle so formed is given by

$$\Delta = \frac{1}{2}\left[\frac{a}{m_1 m_2}\left(\frac{a}{m_2} + \frac{a}{m_3} - \frac{a}{m_3} - \frac{a}{m_1}\right) + \frac{a}{m_2 m_3}\left(\frac{a}{m_3} + \frac{a}{m_1} - \frac{a}{m_1} - \frac{a}{m_2}\right) + \frac{a}{m_3 m_1}\left(\frac{a}{m_1} + \frac{a}{m_2} - \frac{a}{m_2} - \frac{a}{m_3}\right)\right]$$

$$= \frac{1}{2}a^2\left[\frac{a}{m_1 m_2}\left(\frac{a}{m_2} - \frac{a}{m_1}\right) + \frac{a}{m_2 m_3}\left(\frac{a}{m_3} - \frac{a}{m_2}\right) \frac{a}{m_3 m_1}\left(\frac{a}{m_1} - \frac{a}{m_3}\right)\right]$$

Factorizing by cyclic-order method,

$$\Delta = \frac{1}{2}a^2\left(\frac{1}{m_2} - \frac{1}{m_1}\right)\left(\frac{1}{m_3} - \frac{1}{m_2}\right)\left(\frac{1}{m_1} - \frac{1}{m_3}\right) \qquad ...(4)$$

As by hypothesis, m_1, m_2 and m_3 are in H.P.

$\Rightarrow$ $\quad \frac{1}{m_1}, \frac{1}{m_2}, \frac{1}{m_2}$ are in A.P.

$\Rightarrow$ $\quad \frac{1}{m_2} - \frac{1}{m_1} = \frac{1}{m_3} - \frac{1}{m_2} = d$, (say) hence $\frac{1}{m_3} - \frac{1}{m_1} = 2d$

Putting the values in (4), we get

$\Delta = -\ 1/2.\ a^2\ d.\ d\ (-2d) = a^2\ d^3$ which is constant quantity as far as a and d are constant.

Example 85(c):

If a normal to a parabola make an angle ϕ with the axis, show that it will cut the curve again at an angle $\tan^{-1}$ (1/2) $\tan \phi$.

Solution:

Let the equation of the normal to any parabola

$$y^2 = 4ax \qquad ...(1)$$

be $$y = mx - 2am - am^3 \qquad ...(2)$$

So get the other point of contact, we solve (1) and (2) simultaneously.

Putting the value of x from (1) in (2), we get

$$y = \frac{y^2}{4a} - 2am - am^3$$

$$\Rightarrow \quad my^2 - 4ay - 2am - am^3 = 0. \qquad ...(3)$$

As (2) is the normal to (1) at $(am^2, -2am)$, so one point of intersection is $(am^2, -2am)$; hence one factor to (3) is $(y + 2am)$ and therefore the other factor is

$$(my - 4a - 2am^2).$$

So $$y = \frac{1}{m}\left(4a + 2am^2\right).$$

Putting this value in (2),

$$\frac{4a}{m} + 2am = mx - 2am - am^3$$

$$\Rightarrow \quad x = \frac{a}{m^2}\left(2 + m^2\right)^2.$$

So the second point of intersection, say Q, of (1) and (2) is

$$\left[\frac{a}{m^2}\left(2+m^2\right)^2, \frac{2a}{m}\left(2+m^2\right)^2\right]$$

As the equation to a tangent at the point (x_1, y_1) to parabola (1) is $yy_1 = 2a\ (x + x_1)$, its slope is $2a/y_1$. Hence slope of the tangent at the second point of intersection is

$$(2a) / \frac{2a}{m}\left(2+m^2\right) = \frac{m}{2+m^2}$$

If the reqd. angle be q, then $\tan\theta = \dfrac{m - \dfrac{m}{2+m^2}}{1 + m.\dfrac{m}{2+m^2}}$

$$= \frac{2m + m^3 - m}{2 + m^2 + m^2} = \frac{m(1+m^2)}{2(1+m^2)} = \frac{m}{2}$$

So $\quad \tan\theta = \dfrac{m}{2}$.

Given that the angle that the line (2) make with x-axis is ϕ, then

$$m = \tan\phi.$$

Hence $\quad \tan\theta = \dfrac{1}{2}.\tan\theta$

or $\quad \theta = \tan^{-1}\left[\dfrac{1}{2}.\tan\phi\right]$ **Proved.**

Example 86:

What is the equation to the chord of the parabola $y^2 = 8x$ which is bisected at the point (2, – 3)

Solution:

Equation to the parabola is given as $y^2 = 8x$. ...(1)

Any line passing through (2, – 3) may be given by

$$y + 3 = m(x - 2). \quad ...(2)$$

Let (2) cut the parabola (1) at A and B whose co-ordinates are (x_1, y_1) and (x_2, y_2) respectively.

Putting the values of x from (2) in (1), we get

$$y^2 = 8\left[\frac{1}{m}(y+3+2m)\right]$$

or $\quad my^2 - 8y - 8(2m + 3) = 0.$ ...(3)

The ordinates of the points A and B i.e. y_1 and y_2 will be given by (3); so $y_1 + y_2 = 8/m$.

The ordinate of the middle point of AB will be $\dfrac{y_1 + y_2}{2} = \dfrac{4}{m}$ and if (2, – 3) be the mid-point of the chord, we have

$$\frac{4}{m} = -3 \quad \text{or} \quad m = -\frac{4}{3}.$$

Putting the value in (2), we get the required equation as

$$y + 3 = -\frac{4}{3}(x-2)$$

or $4x + 3y + 1 = 0.$ **Ans.**

Example 87:

Prove that the area of the triangle formed by the tangents from the point (x_1, y_1) and the chord of contact is $y_1^2 - 4ax_1)^{3\,2} / 2a$.

Solution:

Let P be the point (x_1, y_1) outside the parabola

$$y^2 = 4ax, \quad ...(1)$$

A and B the points of contact of the tangents from P on (1), as (α, β) and (α', β') respectively, then the equation of the chord of contact AB of the point P will be

$$yy_1 = 2a\,(x + x_1). \quad ...(2)$$

Length of PN, the perpendicular from P on AB will be

$$\frac{2a(x_1+x_2)-y_1 y_1}{\sqrt{(y_1^2+4a^2)}} = \frac{y_1^2-4ax_1}{\sqrt{(y_1^2+4a^2)}} \text{ (omitting-ve sign.)}$$

$$\text{Area of the triangle PAB} = \frac{1}{2}AB \times PN$$

$$= \frac{1}{2}\frac{\sqrt{(y_1^2+4a^2)}\sqrt{(y_1^2+4ax_1)}}{a} \times \frac{\sqrt{(y_1^2+4ax_1)}}{\sqrt{(y_1^2+4a^2)}}$$

[as $AB = \sqrt{(y_1^2 + 4a^2)}\ \sqrt{(y_1^2 + 4ax_1)}/a$, by the previous question]

$$= \frac{(y_1^2-4ax)^{3/2}}{2a}$$

Example 88:

A chord of a parabola passes through a point on the axis (outside the parabola) whose distance from the vertex is half the latus rectum; prove that the normals at its extremities meet on the curve.

Solution:

Let the equation of the parabola be $y^2 = 4ax$.

The point on the axis outside the parabola at a distance of half the latus rectum from the vertex will be $(-2a, 0)$ and let T_1 and T_2 be two points on the parabola as $(at_1^2, 2at_1)$ and $(at_2^2, 2at_2)$ respectively.

The line passing through T_1 and T_2 will be

$$y - 2at_1 = \frac{2at_1 - 2at_2}{at_1^2 - at_2^2}\left(a - at_1^2\right) \text{ or } y - 2at_1 = \frac{2}{t_1 + t_2}\left(x - at_1^2\right)$$

If this line passes through (– 2a, 0), we have

$$0 - 2at_1 = \frac{2}{t_1 + t_2}\left(-2a - at_1^2\right)$$

or $\quad t_1 t_2 = 2$ ...(1)

Equation of the normal at T_1 is

$$y = -t_1x + 2at_1 + at_1^3 \quad ...(2)$$

and equation of the normal at T_2 is

$$y = -t_2x + 2at_2 + at_2^3 \quad ...(3)$$

To solve (2) and (3), subtracting (2) from (3), we get

$$0 = (t_1 - t_2)\ x - 2a\ (t_1 - t_2) - a\ (t_1^3 - t_2^3)$$

$$\Rightarrow \quad x = 2a + a\ (t_1^2 + t_1t_2^2 + t_1^2 \text{ as } t_1 - t_1 \neq 0.$$

Substituting this value of x in (2), we get

$$y = -\ [2a + a\ (t_1^2 + t_1t_2 + t_2^2) + 2at_1 + at_1^3$$
$$= -\ at_1t_2\ (t_1 + t_2).$$

Hence the point of intersection of the normals given by (2) and (3) is

$$[2a + a\ (t_1^2 + t_1t_2 + t_2^2),\ -\ at_1t_1\ (t_1 + t_2)].$$

The equation of the parabola is $y^2 - 4ax = 0$.

Substituting the co-ordinates of the point of intersection in the equation of the parabola, the L. H. S. becomes

$$[-\ at_1t_2\ (t_1 + t_2)]^2 - 4\ [2a + a\ (t_1^2 + t_1t_2 + t_2^2)]$$
$$4a^2\ \{t_1^2 + t_2^2\ 2t_1t_1\} - 4a^3\ (2 + t_1^2 + t_1t_2 + t_2^2\}$$

$$[\because t_1t_2 = 2 \text{by} (1)]$$

$$4a^2\ \{t_1^2 + t_2^2 + 4 - 2 - t_1^2 - 2 - t_2^2] = 0 \text{ R. H. S.}$$

As the point of intersection satisfies the equation of the parabola, it lies on the curve.

Example 89:

Prove that the length of the intercept on the normal at the point (at^2, 2at) made by the circle which is described on the focal distance of the given point as diameter is a √(1 + t2).

Solution:

The equation of the parabola is $\quad y^2 = 4ax$ ...(1)

If the point $(at^2, 2at)$ be P, then the normal at P is

$$y = -tx + 2at + at^2 \quad ...(2)$$

If PS is the diameter of ay circle,

□∠PNS = 90° (being in a semi-circle).

The length of the intercept of the normal by the circle

$$= PN = \sqrt{(PS^2 - SN^2)}. \quad ...(3)$$

Now $PS = \sqrt{(a - at^2)^2 + (0 - 2at)^2}$

$$= a(1 + t^2) \quad ...(4)$$

SN = perpendicular distance of the focus (a, 0) from the normal (2)

$$= \frac{-at + 2at + at^3}{\sqrt{(1+t^2)}} = \frac{at\,1+t^2}{\sqrt{(1+t^2)}} = at\sqrt{(1+t^2)} \quad ...(5)$$

Putting the values in (3) from (4) and (5), we get

$$PN = \sqrt{\{a^2(1 + t^2)^2 - \{a^2t^2(1 + t^2)\}\}}$$

$$= \sqrt{a^2(1 + t^2)(1 + ts^2 - t^2)} = a\sqrt{(1 + t^2)}.$$ **Proved.**

Example 90:

If a perpendicular be let fall from any point P upon its polar prove that the distance of the foot of this perpendicular from the focus is equal to the distance of the point P form the directrix.

Solution:

Let the equation of the parabola be $y^2 = 4ax$. ...(1)

Its focus will be (a, 0) and directrix will be

$$x + a = 0 \quad ...(2)$$

If P is any point (x_1, y_1), the equation of its polar to (1), will be

$$yy_1 = 2a(x + x_1) \quad ...(3)$$

Equation to the line passing through P and perpendicular to (3), will be

$$y - y_1 = -\frac{y_1}{2a}(x - x_1)$$

or $\quad 2ay = -y_1x + x_1y_1 + 2ay_1.$ ...(4)

Solving these equations, we get

$$x = -\frac{x_1 y_2^2 + 2ay_1^2 - 4a^2 x_1}{y_1^2 + 4a^2} \text{ and } y = \frac{4ay_1(x_2 + a)}{y_1^2 + 4a^2}$$

So the co-ordinates of the point of intersection of (3) and (4) i.e. the foot of the perpendicular from P on the polar, say N, are

$$\left[\frac{x_1 y_2^2 + 2ay_1^2 - 4a^2 x_1}{y_1^2 + 4a^2}, \frac{4ay_1(x_1 + a)}{y_1^2 + 4a^2}\right]$$

Distance between the focus (a, 0) and N is given by

$$\sqrt{\left[\left\{\frac{x_1 y_2^2 + 2ay_1^2 - 4a^2 x_1}{y_1^2 + 4a^2} - a\right\}^2 + \left(\frac{4ay_1(x_1 + a)}{y_1^2 + 4a^2} - 0\right)^2\right]}$$

$$= \frac{1}{y_1^2 + 4a^2}\sqrt{\left[\left(x_1 y_1^2 + ay_1^2 - 4a^2 x_1 - 4a^3\right)^2 + 16a^2 y_1^2 (x_1 + a)^2\right]}$$

$$= \frac{1}{y_1^2 + 4a^2}\sqrt{\left[\left(y_1^2 - 4a^2\right)^2 (x_1 + a)^2 + 16a^2 y_1^2 (x_1 + a)^2\right]}$$

$$= \frac{(x_1 + a)}{y_1^2 + 4a^2}\sqrt{\left(y_1^2 - 4a^2\right)^2 + 16a^2 y_1^2} = \frac{(x_1 + a)}{\left(y_1^2 + 4a^2\right)} \times \left(y_1^2 + 4a^2\right)$$

$$= x_1 + a \qquad \text{...(5)}$$

Distance of the point P (x_1, y_1) from the directrix $x + a = 0$ is clearly $(x_1 + a)$ which is equal to the distance given by (5). **Hence proved.**

Example 91(a):

Prove that the length of the chord joining the points of contact of tangents drawn from the point (x_1, y_1) is

$$\frac{\sqrt{y_1^2 + 4a^2}\sqrt{y_1^2 - 4a_1}}{a}.$$

Solution:

Let the equation to the parabola by $y^2 = 4ax$. ...(1)

and (x_1, y_1) be any point outside it.

The equation to the chord or contact from (x_1, y_1) to (1) is

$$yy_1 = 2a(x + x_1) \quad \text{or} \quad x = \frac{1}{2a}(yy_1 - 2ax_1) \qquad \text{...(2)}$$

To find out the point of intersection, we solve (1) and (2)

$$y^2 = 4a\frac{1}{2a}(yy_1 - 2ax_1)$$

$\Rightarrow \qquad y^2 - 2yy_1 + 4ax_1 = 0.$ ...(3)

Let the co-ordinates of the points of intersection of (1) and (2) be (α, β) and (α', β'). So b and b' will be given by (3). So $\beta + \beta' = 2y_1$ ...(4)

and $\qquad \beta\beta' = 4ax_1.$ ...(5)

Again as (α, β) and (α', β') be on (2), the points will satisfy it. So

$$\alpha = \frac{1}{2a}(y_1\beta - 2ax_1)$$

$$\Rightarrow \qquad \alpha' = \frac{1}{2a}(y_1\beta' - 2ax_1)$$

Subtracting one from the other, we get

$$\alpha - \alpha' = \frac{y_1}{2a}(\beta - \beta'). \qquad ...(6)$$

The distance between the points (α, β) and (α', β') is

$$\sqrt{\{(\alpha - \alpha')^2 + (\beta - \beta')^2\}} = \sqrt{\frac{y_1^2}{4a^2}(\beta - \beta')^2 + (\beta - \beta')^2}$$

Putting the value of $(\alpha, - \alpha')$ from (6),

$$= \sqrt{\left[(\beta - \beta')^2\left(\frac{y_1^2}{4a^2} + 1\right)\right]}$$

$$= \frac{1}{2a}\sqrt{(y_1^2 + 4a^2)}\sqrt{\{(\beta + \beta')^2 - 4\beta\beta'\}}$$

$$= \frac{1}{2a}\sqrt{(y_1^2 + 4a^2)}\sqrt{\{(2y_1)^2 + 4.4ax\}}$$

Putting the values from (5) and (4) we have

$$= \frac{1}{2a}\sqrt{(y_1^2 + 4a^2) \times 4(y_1^2 - 4ax_1)}$$

$$= \frac{\sqrt{(y_1^2 + 4a^2)} \cdot \sqrt{(y_1^2 - 4ax)}}{a}$$

Proved.

Example 91(b):

P, Q and R three points on a parabola and the chord PQ cuts the diameter through R in V. Ordinates Pm and QN are drawn to this diameter. Prove that RM.RN = RV².

Solution:

Let the equation of the parabola referred to the diameter through R and the tangent at R as axis be $y^2 = 4px$ and let the points P and Q be $(pt_1^2, 2pt_1)$ and $(pt_2^2, 2pt_2)$ respectively.

Equation of the diameter RN is $y = 0$ and the ordinates through P and Q are $x = pt_1^2$ and $x = p_2t^2$. ...(1)

Hence $RM.RN = p_2t^2 . pt_2^2 = (pt_1t_2)^2$.

Also the equation of PQ is

$$y - 2pt_1 = \frac{2pt_2 - 2pt_1}{pt_2^2 - pt_1^2}\left(x - pt_1^2\right)$$

or $y(t_1 + t_2) = 2x + 2pt_1t_2$ ...(2)

It meets x-axis i.e. $y = 0$ at L. So putting $y = 0$ in (2), $x = -pt_1t_2$; hence the co-ordinates of L are $(-pt_1t_2, 0)$.

So $RL = -pt_1t_2$

or $RL^2 = (pt_1t_2)^2$. ...(3)

By (1) and (3) we get $RN.RM = RL^2$. **Proved.**

Example 92:

A chord is a normal to a parabola and is inclined at an angle θ to the axis, prove that the area of the triangle formed by it and the tangents at its extremities is $4a^2 \sec^3 \theta \operatorname{cosec}^3 \theta$.

Solution:

Let the extremities of the normal chord be P and Q and the tangents at P and Q to the parabola say $y^2 = 4ax$ meet in T. Let the co-ordinates of T be (x_1, y_2).

PQ will be the chord of contact for T with respect to the parabola so area of triangle TPQ will be

$$[y_1^2 - 4ax_1]^{3/2}/2a \qquad ...(1)$$

The equation to the chord of contact of T will be

$$yy_1 = 2a(x + x_1)$$

$$\Rightarrow \qquad y = \frac{2ax}{y_1} + \frac{2ax_1}{y_1}. \qquad ...(2)$$

Equation to any normal to $y^2 = 4ax$ is

$$y = mx - 2am - am^2. \qquad ...(3)$$

So (2) and (3) must be identical. As the coefficients of y are equal, others must also equal, so

$$m=\frac{2a}{y_1} \quad \text{and} \quad -2am - am^3 = \frac{2ax_1}{y_1}.$$

whence $y_1 = \frac{2a}{m}$ and $x_1 = \left(-2a - am^2\right)$.

If the inclination of the chord of contact, i.e., the normal is θ; then

$$m = \tan \theta.$$

So $y_1 = \frac{2a}{m} = 2a\cot\theta$ and $x_1 = \left(-2a - am^2\,\theta\right)$.

Substituting in (1), we get

Area of the triangle = $[4a^2 \cot^2 \theta - 4a\,(-2a - a\tan^2 \theta)]^{3/2}/2a$

$$= \frac{1}{2a}\left[4a^2\cot^2\theta + 8a^2\,4a^2\tan^2\theta\right]^{3/2} = \frac{1}{2a}\left[4a^2\left(\cot^2\theta + 2 + \tan^2\theta\right)\right]^{3/2}$$

$$= \frac{1}{2a}\left(4a^2\right)^{3/2}\left\{(\cot\theta + \tan\theta)^2\right\}^{3/2}$$

$$= \frac{1}{2a}8a^3\left\{(\sec\theta + \operatorname{cosec}\theta)^2\right\}^{3/2}$$

$$\left[\because \cot\theta + \tan\theta = \frac{\cos^2\theta + \sin^2\theta}{\sin\theta.\cos\theta} = \sec\theta.\operatorname{cosec}\theta\right]$$

$= 4a^2 \sec^3 \theta \,.\, \operatorname{cosec}^3 \theta.$ **Proved.**

Example 93:

Parabolas are drawn to touch two given straight lines which are inclined at an angle ω; if the chords of contact all pass through a fixed point, prove that (1) their directories all pass through another fixed point, and (2) their facial all lie on a circle which goes through the intersection of the two given straight lines.

Solution:

The equation of the directrix is written by

$$\frac{x + y\cos\omega}{a} + \frac{y + x\cos\omega}{b} = 1 \qquad \text{...(1)}$$

and as the chord of contact passes through a fixed point whose co-ordinates are (h, k), we obtain

$$\frac{h}{a} + \frac{k}{b} = 1.$$

Therefore the directrix passes through the fixed point given by the following equation

$$x + y \cos \omega = h \text{ and } y + x \sec \omega = k.$$

And the focus is given by $ax = by = x^2 + y^2 + 2xy \cos \omega$, whence

$$a = \frac{x^2 + y^2 + 2\,xy \cos \omega}{x}$$

and
$$b = \frac{x^2 + y^2 + 2\,xy \cos \omega}{y}.$$

Substituting for 'a' and 'b' in (2) and simplifying, we have $x^2 + y^2 + 2xy \cos \omega - xh - yk = 0$, which is a circle which pass through the origin.

Example 94:

Prove that the sum of the angles which the three normals, drawn from any point O, make with the axis exceeds the angle which the focal distance of O makes with the axis by a multiple of π.

Solution:

Let $y = mx - 2am - am^3$ be the normal passing through O, whose co-ordinates are (h, k) (say); then $k = mh - 2am - am^3$ or $am^3 + m(2a - h) + k = 0$.

Then we have

$$m_1 + m_2 + m_3 = 0,\ \Sigma m_1 m_2$$

$$= \frac{2a - h}{a} \text{ and } m_1 m_2 m_3$$

$$= \frac{-k}{a}.$$

If θ_1, θ_2 and θ_3 be the angles of inclination of these normals with axis, *i.e.*, x-axis then $m_1 = \tan \theta_1$, $m_2 = \tan \theta_2$ and $m_3 = \tan \theta_3$.

If S is the focus (a, 0), then the slope of $SO = \dfrac{k}{h - a} = \tan \alpha$ (say)

Again $\tan(\theta_1 + \theta_2 + \theta_3)$

$$= \frac{(m_1 + m_2 + m_3 - m_2\ m_2\ m_3)}{1 - (m_1\ m_2 + m_2\ m_3 + m_3\ m_1)} = \frac{0 - (-k/a)}{1 - \left(\dfrac{2a - h}{a}\right)} - \frac{k}{h - a}$$

$$\Rightarrow \qquad \tan(\theta_1 + \theta_2 + \theta_3) = \tan \alpha.$$

Hence $\theta_1 + \theta_2 + \theta_3 = n\pi + \alpha.$

Example 95(a):

If P be fixed, then QR is fixed in direction and the locus of the centre of the circle circumscribe PQR is a straight line.

Solution:

The slope of QR

$$= \frac{-2a\left[m_2 - m_3\right]}{a\left[m_2^2 - m_3^2\right]} = \frac{-2}{m_2 + m_3} = \frac{2}{m_1}$$

If m_1 is constant $2/m_1$ is also constant, hence the direction QR is fixed. The centre of the circle given

$$\left(-g, -f\right) \text{ i.e., } \left(\frac{h+2a}{a}, \frac{k}{4}\right).$$

So if (x, y) is the centre, $x = \dfrac{h+2a}{a}$ ...(1)

and $y = \dfrac{k}{4}$ ...(2)

Again the equation of normal at $P \equiv \left(am_1^2 - 2am_1\right)$ is

$$y = m_1x - 2am_1 - am_1^3. \quad ...(3)$$

As this passes through (h, k) and also m_1 = constant, then $k = m_1h - 2am_1 - am_1^3$. Eliminating h and k from (1), (2) and (3), we have $4y = m_1(2x - 2a) - 2am_1 - am_1^3$. This is the locus of centre (xm y) and it is clearly a straight line.

Example 95(b):

Prove that the equation to the circle, which power through the focus and touches the parabola $y^2 = 4ax$ at the point $(at^2, 2at)$, is $x^2 + y^2 - ax(3t^2 + 1) - ay(3t - t^3) + 3a^2t^2 = 0$.

Prove also that the locus of its centre is the curve

$$27ay^2 = (2x - a)(x - 5a)^2.$$

Solution:

If the circle touches the parabola $y^2 = 4ax$ at $(at^2, 2at)$, they must have a common tangent at that point, and hence a common normal. The centre of the circle must lie on that normal. Let (h, k) be the co-ordinates of the centre of the centre and r be the radius of the circle. Then its equation is $\quad x^2 + y^2 - 2hx - 2ky + c = 0$ ...(1)

The equation to the normal at $(at^2, 2at)$ is

$$y = -tx + 2at + at^3 \quad ...(2)$$

As the centre (h, k) lies on the normal (2), hence

$$k = -th + 2at + at^3. \quad ...(3)$$

Focus of the parabola is (a, 0).

As the circle passes through (a, 0) and $(at^2, 2at)$, the distance of these points from the centre (h, k) must each be equal to the radius. So we have

$$r^2 = (h - a)^2 + k^2 = (h - at^2)^2 + (k - 2at)^2$$

$$\Rightarrow \quad -2ah + a^2 = -2aht^2 - 4akt + a^2t^4 + 4a^2t^2$$

$$\Rightarrow \quad 4kt = h(1 - t^2) + at^4 + 4at^2 - a \quad ...(4)$$

Solving (3) and (4), $2h = a(3t^2 + 1)$ ...(5)

and $\quad 2k = a(3t - t^3)$ ...(6)

Putting the values of h and k in (1), the equation to the circle is given by

$$x^2 + y^2 - a(3t^2 + 1)x - a(3t - t^3)y + c = 0 \quad ...(7)$$

As (7) passes through the focus (a, 0), the co-ordinates will satisfy it:

Hence we have $a^2 - a^2(3t^2 + 1) + c = 0$;

$\therefore \quad c = 3a^2t^2.$

Putting the values in (7) the circle given by

$$x^2 + y^2 - ax(3t^2 + 1) - ay(3t - t^3) + 3a^2t^2 = 0.$$

(ii) To get the locus of the centre, we have to eliminate t from (5) and (6).

Multiplying (5) by t and (6) by 3 and adding, we get $2th + 6k = 10at$; hence $t = 3k/(5a - h)$. Substituting the value of 't' in (5), we have $(2h - a)$

$$= 3a.\left(\frac{3k}{5a - h}\right)^2.$$

Simplifying and generalising, the locus of centre (h, k) is

$$(2x - a)(x - 5a)^2 = 27ay^2.$$ **Hence proceed.**

Example 95(c):

Three normals are drawn to the parabola $y^2 = 4ax \cos \alpha$ from any point lying on the straight line $y = b \sin \alpha$. Prove that the locus of the orthocentre of the triangles formed by the corresponding tangents is the curve $\frac{x^2}{a^2} + \frac{y^2}{b^2} = 1$, the angle α being variable.

Solution:

The equation to the parabola is given as

$y^2 = 4ax \cos \alpha$ (i)

Equation to any normal to (i) is $y = mx - 2am \cos \alpha - am^3 \cos \alpha$.

If this normal passes through any point on $y = b\ \alpha$. Then we have

$b \sin \alpha = mx - 2am \cos \alpha - am^3 \cos \alpha$

$\Rightarrow \quad am^3 \cos \alpha + 2am \cos \alpha - mx + b \sin \alpha = 0.$ (ii)

Let m_1, m_2 and m_3 be its roots; then we have

$$m_1 + m_2 + m_3 = 0,\ m_1 m_2 m_3 = \frac{b \sin \alpha}{a \cos \alpha} \text{ and } \Sigma m_1 m_2 = \frac{2a \cos \alpha - x}{a \cos \alpha}$$

The feet of the perpendicular of the normals are given as

$$\left(am_1^2 \cos \alpha, -2am_1 \cos \alpha\right);\ \left(am_2^2 \cos \alpha, -2am_2 \cos \alpha\right)$$

and $\left(am_3^2 \cos \alpha, -2am_3 \cos \alpha\right)$. Now the equation to the tangent at $\left(am_1^2 \cos \alpha, -2am_1 \cos \alpha\right)$ w.r.t. the parabola is

$$y\ (-2am_1 \cos \alpha) = 2a \cos \alpha \left(x + am_1^2 \cos \alpha\right)$$

$$\Rightarrow \quad -ym_1 = x + am_1^2 \cos \alpha. \qquad ...(3)$$

Similarly tangent at $\left(am_2^2 \cos \alpha, -2am_2 \cos \alpha\right)$ is given by

$$-ym^2 = x + am_2^2 \cos \alpha.$$

Solving the two, the co-ordinates of the point of intersection of the tangents are $\left\{am_1 m_2 \cos \alpha, -a\ (m_1 + m_2) \cos \alpha\right\}$

or $\left(\frac{b \sin \alpha}{m_3} . am_3 \cos \alpha\right)$ [on putting the values from (2)].

Similarly the other two vertices of the triangles formed by the tangents are given as

$$\left(\frac{b \sin \alpha}{m_2}, am_2 \cos \alpha\right) \text{ and } \left(\frac{b \sin \alpha}{m_1}, am_1 \cos \alpha\right)$$

The equation of the line perpendicular of line (3) through the vertex $\left(\frac{b \sin \alpha}{m_1}, am_1 \cos \alpha\right)$ is given by

$$y = m_1 x + am_1 \cos \alpha - b \sin \alpha \qquad ...(4)$$

Similarly other perpendicular line is given as

$$y = m_2x + am_2 \cos\alpha - b \sin\alpha. \qquad ...(5)$$

On solving (3) and (4), we get

$$x = -a \cos\alpha, \qquad ...(6)$$

$$y = -b \sin\alpha. \qquad ...(7)$$

The ortho-centre of the given circle is (x, y)

Hence eliminating α from (6) and (7), we get the required focus as

$$\cos^2\alpha + \sin^2\alpha = \frac{x^2}{a^2} + \frac{y^2}{b^2} = 1.$$

Example 96:

If the normals at the three points P, Q and R meet in a point and if PP', QQ' and RR' be chords parallel to QR, RP and PQ respectively, prove that the normals at P', Q', R' also meet in a point.

Solution:

Let P, Q and R be $\left(am_1^2, -2am_1\right)$, $\left(am_2^2, -2am_2\right)$ and $\left(am_3^2, -2am_3\right)$ respectively and P', Q' and R' be $\left(at_1^2, -2at_1\right)$, $\left(at_2^2, -2at_2\right)$ and $\left(at_3^2, 2at_3\right)$ respectively.

Then Slope of $PP' = \dfrac{-2am_1 + 2at_1}{am_1^2 - at_1^2} = \dfrac{-2}{m_1 + t_1}$

and Slope of $QR = \dfrac{-2am_2 + 2am_3}{am_1^2 - am_3^2} = \dfrac{-2}{m_2 + m_3}$

As PP' as parallel to QR, slopes must be equal.

Hence $m_1 + t_1 = m_2 + m_3 = -m_1$ $\qquad \{\because \Sigma m_1 = 0\}$

$\therefore \quad t_1 = -2m_1.$

Similarly $t_2 = -2m_2$ and $t_3 = -2m_1 - 2m_2 - 2m_3.$

$= -2(m_1 + m_2 + m_3) = 0$ (as $\Sigma m_1 = 0$) or $\Sigma t_1 = 0.$

Hence the normals at point P', Q' and R' meet in a point.

Example 97:

Show that three circles can be drawn to touch a parabola and also to touch at the focus a given straight line passing through the focus, and prove that the tangents at the point of contact with the parabola form an equilateral triangle.

Solution:

We know that the equation to the circle which passes through the focus (a, 0) and also touches the parabola $y^2 = 4ax$ at $(at^2, 2at)$ is $x^2 + y^2 - ax\,(3t^2$

+ 1) – ay $(3t - t^3) + 3a^2t^2 = 0$ and equation of any line is $x^2 + y^2 - ax(3t^2 + 1) - ay(3t - t^3) + 3a^2t^2 = 0$ and equation of any line through (a, 0) is given by $y = \frac{\sin\alpha}{\cos\alpha}(x-a)$ $\left(\because \quad m = \tan\alpha = \frac{\sin\alpha}{\cos\alpha}\right)$

$x \sin\alpha - y\cos\alpha = a\sin\alpha$ and equation of perpendicular line will be

$$x\cos\alpha + y\sin\alpha = a\cos\alpha \qquad ...(1)$$

Co-ordinates of the centre of the circle given are $\{a(3t^2 + 1)/2,\ a(3t - t^3)/2\}$.

Since this passes through the centre then we have

$$\frac{a(4t^2+1)}{2}\cos\alpha + \frac{a}{2}(3t-t^3)\sin\alpha = a\cos\alpha$$

$$\Rightarrow \quad (3t^2 + 1)\cos\alpha + (3t - t^3)\sin\alpha = 2\cos\alpha$$

$$\Rightarrow \quad (3t - t^3)\sin\alpha = (1 - 3t^2)\cos\alpha$$

$$\therefore \quad \cot\alpha = \frac{3t-t^3}{1-3t^2}.$$

Putting $t = \tan\theta$, we get $\cot\alpha = \dfrac{3\tan\theta - \tan^3\theta}{1-3\tan^2\theta} = \tan 3\theta$

$$\Rightarrow \quad \tan 3\theta = \tan\left(\frac{\pi}{2}-\alpha\right)$$

then we have

$$3\theta = \left(\frac{\pi}{2}-\alpha\right), \left\{\pi + \left(\frac{\pi}{2}-\alpha\right)\right\} \text{ or } \left\{2\pi + \left(\frac{\pi}{2}-\alpha\right)\right\}$$

If θ_1, θ_2 and θ_3 be three values of θ, then

$$\theta_1 = \frac{1}{3}\left(\frac{\pi}{2}-\alpha\right), \theta_2 = \frac{\pi}{3} + \frac{1}{3}\left(\frac{\pi}{2}-\alpha\right) \text{ and } \theta_3 = \frac{2\pi}{3} + \frac{1}{3}\left(\frac{\pi}{2}-\alpha\right)$$

So $\qquad \theta_2 - \theta_1 = \dfrac{\pi}{3} = \theta_3 - \theta_2.$

Therefore the three normals are inclined at 60°. So the angles between the tangents are also 60°.

Therefore these form an equilateral triangle.

Example 98:

Two of the normals drawn from a point O to the curve know complementary angles with the axis; prove that the locus of O and the curve which is touched by its polar are parabolas such that their lateral recta and that of the original parabola form a geometrical progression. Sketch the three curves.

Solution:

Let the equation of the parabola be $y^2 = 4ax$ as the normal $y = mx - 2am - am^3$ passes through O, whose co-ordinates are

(h, k), we have $m_1 + m_2 + m_3 = 0,$ ...(1)

$$\Sigma m_1 m_2 = \frac{ah - h}{a}$$

and $\Sigma m_1 m_2 m_3 = -\frac{k}{a}.$ ...(3)

Let the angles which the normals having slopes m_1 and m_2 make with x-axis be θ and ϕ; then $\theta + \phi = 90^\circ$ or $\phi = (90^\circ - \theta)$.

So $m_1 = \tan \theta_1$ and $m_2 = \tan (90^\circ - \theta) = \cot \theta$

$\therefore$ $m_1 m_2 = 1.$ ...(4)

Again $m_3 (m_2 + m_1) + m_1 m_2 = \frac{2a - h}{a}$ [by (2)]

But $m_1 + m_2 = -m_3$ [by (1)]

So $-m_3^2 + 1 = \frac{2a - h}{a}$ or $-\frac{k^2}{a^2} = \frac{2a - h}{a} - 1 = \frac{a - h}{a}$

$\Rightarrow$ $k^2 = a (h - a).$ ...(5)

Generalising, we obtain the locus of O as $y^2 = a (x - a)$ which as clearly a parabola having latus rectum equal to a.

Again the equation of the polar of (h, k) w.r.t., $y^2 = 4ax$ is

given by $yk = 2a (x + h) = 2a \; x + \left(\frac{k^2 + a^2}{a}\right)$ from (5)

$$\Rightarrow \quad y = \frac{2a}{k}(x + a) + 2k$$

$$\Rightarrow \quad y = \frac{2a}{k}(x + a) + \frac{4a}{2a/k}.$$

Changing the origin to (–a, 0), this becomes $y = \frac{2a}{k} x + \frac{4a}{2a/k}$ which is of the form $y = mx + 4a/m$, and hence is tangent to $y^2 = 4.4ax$.

Changing the origin back to the original one, parabola becomes

$$y^2 = 16a (x + a).$$

Hence the polar of (h, k) touches a parabola $y^2 = 16a (x + a)$.

Latus rectum of this parabola is 16a.

Clearly a, 4a, 16a are in G.P. **Hence proved.**

Example 99:

If the normals drawn from any power line x = 2a in points whose ordinates are in arithmetical progression, prove that the tangents of the angles which the normals make with the axis are in geometrical progression.

Solution:

Let the equation of the parabola be $y^2 = 4ax$. If the feet of the normals passing through a point be $\left(am_1^2, -2am_1\right)$, $\left(am_2^2, -2am_2\right)$ and $\left(am_3^2, -2am_3\right)$, equation of normal at $\left(am_1^2, -2am_1\right)$ is $y = m_1x - 2am_1 - am_1^3$.

Solving with the line x = 2a, we get the point of intersection as $(2a, - am^3)$. Similarly solving with the other normals the ordinates of the points of intersection with x = 2a are $-am_1^3, -am_2^3$ and $-am_3^3$.

These ordinates are in A.P.

So $$-am_1^3 - am_3^3 = -2am_2^3$$

or $$m_1^3 + m_3^3 = 2m_2^3. \quad ...(1)$$

But $m_1 + m_3 = - m_2$, as $\Sigma m_1 = 0$ ...(2)

By (1) $2m_2^3 = (m_1 + m_3)\left(m_1^2 + m_3^2 - m_1m_3\right)$

$$-2m_2^2 = \left(m_1^2 + m_3^2 - m_1m_3\right) \text{ (as } m_1 + m_3 = - m_2) \quad ...(3)$$

and by (2) $m_2^2 = m_1^2 + m_3^2 + 2m_1m_3$. ...(4)

Subtracting (3) from (4), $3m_2^2 = 3m_1m_3$ or $m_2^2 = m_1m_3$.

Example 100:

PG, the normal at P to a parabola, cuts the axis in G and is produced to Q so that GQ = 1/2PG; prove that the other normals which pass through Q intersect at right angles.

Solution:

Let the co-ordinates of P be $\left(am_1^2, -am_1\right)$.

Equation of normal at P will be given as

$$y = m_1x - 2am_1 - am_1^3. \quad ...(1)$$

Solving (1) with x-axis *i.e.* y = 0, the abscissa of the point of intersection is $(2a + am_1^2)$.

The point Q is such that GQ = 1/2 PG. Hence Q divides PG in the ratio of 3: 1 externally. Let co-ordinates of Q be (h, k); then we have

$$k = \frac{3 \times 0 - 1(-2am_1)}{3-1} = am_1 \text{ or } m_1 = \frac{k}{a}.$$

But $\quad m_1 m_2 m_3 = \frac{-k}{a}$

Putting the value of m_1, we get $\frac{k}{a} m_2 m_3 = \frac{-k}{a}$ or $m_2 m_3 = -1$.

Hence the two normals through Q are perpendicular to each other.

Example 101:

A circle is described whose centre is the vertex and whose diameter is three-quarters of the latus rectum of a parabola; prove that the common chord of the circle and parabola bisects the distance between the vertex and the focus.

Solution:

Let the equation of the parabola be $y^2 = 4ax$

Its vertex is (0, 0) and latus rectum is 4a. Hence the centre of given circle will be (0, 0) and diameter will be 3/4.4a = 3a.
So its equation will

To get the point of intersection of (1) and (2), we solve them simultaneously. So putting the value of y^2 from (1) in (2), we get

$$x^2 + 4ax = \frac{9}{4}a^2$$

$$\Rightarrow \quad 4x^2 + 16ax - 9a^2 = 0$$

$$\Rightarrow \quad (2x - a)(2x + 9a) = 0, \text{ whence } x = a/2.$$

(neglecting –ve value as the parabola does not lie on that side).

As both the curves (1) and (2) are symmetrical about axis of x, the common chord will be parabola to axis of y.

Hence its equation will be x = a/2.

Clearly it cuts x-axis at (a/2, 0) which is mid-way between the vertex (0, 0) and the focus (a, 0).

Example 102:

PR and QR are chords of a parabola which are normals at P and Q. Prove that two of the common chords of the parabola and the circle circumscribing the triangle PRQ meet on the directrix.

Solution:

Suppose the co-ordinates of R on the parabola

$y^2 = 4ax$ be (h, k) and let $h = am_3^2$ and $k = -2am_3$.

Again, let the co-ordinates of P and Q be $(am_1^2, -2am_1)$ and $(am_2^2, -2am_2)$ respectively. Now, if any normal $y = mx - 2am - am^3$ passes through (h, k). Then we have

$$m_1 + m_2 + m_3 = 0,\ m_1 m_2 + m_2 m_3 + m_3 m_1 = \frac{2a-h}{h}$$

and $$m_1 m_2 m_3 = \frac{-k}{a}.$$

But as $m_3 = \frac{-k}{2a}$, so $m_1 m_2 = 2$ and $m_1 + m_2 = \frac{k}{2a}$. ...(1)

We have proved that the circle and parabola intersect in four points and that the line joining one pair of these four points and the line joining the other pair are equally inclined to the axis. Therefore, if S be the fourth point, then RQ and RS are equally inclined with the axis.

Again, the equation to PQ will be $-y(m_1 + m_2) = 2x + 2am_1m_2$

$$\Rightarrow \quad -y\frac{k}{2a} = 2x + 4a \quad \text{[from (1)]}$$

$$\Rightarrow \quad -ky = 4ax + 8a^2. \quad ...(2)$$

The equation to RS is $k(y - k) = 4a(x - h)$ or $yk = 4ax + k^2 - 4ah$.

But $k^2 = 4ah$. So $yk = 4ax$. ...(3)

Adding (3) with (2), we have $8axk + 8a^2 = 0$ or $x + a = 0$. This is the equation of directrix. Hence the common chords PR and RS of parabola and circle meet in directrix.

Example 103:

Prove that the locus of the centre of the circle, which passes through the vertex of a parabola and through its intersections with a normal chord, is the parabola $2y^2 = ax - a^2$.

Solution:

Equation of normal at point $(at_1^2, 2at_1)$ of the parabola $y^2 = 4ax$ is $y = -t_1x + 2at_1 +$, this line intersects the parabola at point $(at_2^2, 2at_2)$, such that $t_2 + t_1 = -2/t_1$

The equation of any circle passing through (0, 0) is

$$x^2 + y^2 - 2gx - 2fy = 0.$$

If this passes through $(at_1^2, 2at_1)$, then

$$at_1^3 + 4at_1 - 2gt_1 - 4f = 0 \quad ...(2)$$

Similarly, if the circle passes through $\left(at_2^2, 2at_2\right)$, then we have

$$at_2^3 + 4at_2 - 2gt_2 - 4f = 0. \quad ...(3)$$

The locus of the centre (g, f) of the circle will be obtained by eliminating t_1, t_2 and t_3 from (1), (2) and (3).

Subtracting (3) from (2), we have

$$a\left(t_1^3 - t_2^3\right) + 4a\left(t_1 - t_2\right) - 2g\left(t_1 - t_2\right) = 0$$

$$\Rightarrow \quad a\left(t_1^2 + t_2^2 + t_1t_2\right) + 4a - 2g = 0$$

$$\Rightarrow \quad a\left(t_2^2 - 2\right) + 4a - 2g = 0 \quad \text{[by (1)]}$$

$$\Rightarrow \quad at_2^2 + 2a - 2g = 0. \quad ...(4)$$

Again multiplying (2) by t_2 and (3) by t_1 and then subtracting, we have

$$at_1t_2\left(t_1^2 - t_2^2\right) + 4f\left(t_1 - t_2\right) = 0$$

$$\Rightarrow \quad at_1t_2\,(t_1 + t_2) + 4f = 0 \quad \text{[by (1)]}$$

$$\Rightarrow \quad -2at_2 + 4f = 0$$

Hence $t_2 = 2f/a$.

Substituting this value of t_2 in (4), we have

$$a\,(4f^2/a^2) + 2a - 2g = 0 \text{ or } 2f^2 + a^2 - ag = 0.$$

Gemeralising, the locus of the centre (g, f) is

$$2y^2 + a^2 - ax = 0 \text{ or } 2y^2 = ax - a^2.$$ **Hence Proceed.**

Example 104(a):

A parabola of latus rectum l, touches a fixed equal parabola, the axes of the two curves being parallel; prove that the locus of the vertex of the moving curve is a parabola of latus rectum 2l.

Solution:

Let (h, k) be the co-ordinates of the vertex of the moving parabola and its equation be $(y - k)^2 + 1\,(x - h) = 0$. The equation of the fixed curve be $y^2 - lx = 0$; then the tangents at the point (lt^3, lt) to the two curves are

$$-2yt + x + lt^2 = 0 \text{ and } (lt - k) + \frac{1}{2}x + \frac{l^2t^2}{2} + k^2 - lh - kl\,t = 0$$

As they are coincident $\dfrac{k - l\,t}{2t} = \dfrac{l/2}{1} = \dfrac{l^2t^2 + 2k - 2lh - 2klt}{2l\,t^2}$

from which $t = \dfrac{k}{2l}$ and $k^2 - l\,h = kl\,t$.

Eliminating t, we get $k^2 = 2l\,h$. Hence the locus of the vertex is $y^2 = 2l\,x$ which is a parabola having latus rectum as $2l$.

Example 104(b):

The two parabolas $y^2 = 4a\,(x - l)$ and $x^2 = 4a\,(y - l)$ always touch one another, the quantities l and l' being both variable; prove that the locus of their point of contact is the curve $xy = 4a^2$.

Solution:

Any point on parabola $y^2 = 4a\,(x - l)$ may be taken as $(l + at_1^2, 2at_1)$ and on x^2 4a $(y - l')$ as $(2at_2, l' + at_2^2)$. Let (x, y) be the common point, then

$$xy = 2at_2 \times 2at_1 \text{ or } xy = 4a^2t_1t_2. \qquad ...(1)$$

Tangent at point 't_1' of the parabola $y^2 = 4a\,(x - l)$ will be

$$t_1y = x + at_1^2 \qquad ...(2)$$

Again the tangent at 't_2' of the parabola $x^2 = 4a\,(y - l')$ will be

$$xt_2 = y + 2at_2 \text{ or } y = xt_2 - 2at_2 \qquad ...(3)$$

The tangents (2) and (3) will coincide, if

$$t_2 = 1/t_1 \text{ or } t_1t_2 = 1. \qquad ...(4)$$

By (1) and (4), we have $xy = 4a^2$, which is the locus of common point of the parabola given.

Example 104(c):

The sides AB and AC of a triangle ABC are given in position and the harmonic mean between the lengths AB and AC is also given; prove that the locus of the focus of the parabola touching the sides at B and C is a circle whose centre lies on the line bisecting the angle BAC.

Solution:

Let AB and AC be the axes and each equal to a and b respectively. Say k is the harmonic mean between a and b; then

$$\frac{1}{a} + \frac{1}{b} = \frac{2}{k} \qquad ...(1)$$

Now the foci are given by $ax = by = x^2 + y^2 + 2ay \cos \omega$ whence

$$a = \frac{x^2 + y^2 + 2xy \cos \omega}{x} \qquad ...(2)$$

and $$b = \frac{x^2 + y^2 + 2xy \cos \omega}{y} \qquad ...(3)$$

Putting the values of a and b from (2) and (3) in (1) and simplifying, we get

$$\frac{2}{k} = \frac{x+y}{x^2 + 2xy\cos\omega + y^2}$$

Hence the locus is $x^2 + 2xy \cos w + y^2 - \frac{k}{2}(x + y) = 0$, and it is clearly a circle.

Again as the abscissa and the ordinate of the centre are equal, so it lies on the bisector of the angle.

Example 104(d):

Prove that the sum of the angles which the four common tangents to a parabola and a circle make with the axis is equal to $n\pi + 2a$, where α is the angle which the radius from the focus to the centre of the circle makes with the axis and n is an integer.

Solution:

Let the equation of parabola be $y^2 = 4ax$. ...(1)

Equation to any tangent to (1) is $y = mx + a/m$. ...(2)

Solving (2) with any circle $x^2 + y^2 - 2fy + c = 0$, we get $x^2(1 + m^2) + 2x(a - g - fm) + a^2/m^2 - 2f(a/m) + c = 0$.

If the line (1) touches the circle, then the roots of this equation must be equal and hence its discriminant will be zero.

So $\quad 4(a - g - fm)^2 = 4(1 + m^2)(a^2/m^2 - 2fm\, a/m + c)$

$\Rightarrow \quad m^4(f^2 - c) + 2gfm^3 - (g^2 - 2ag - c)m^2 + 2fam - a^2 = 0.$

Let its roots be m_1, m_2, m_3 and m_4 and let $m_1 = \tan\theta_1$, $m_2 = \tan\theta_2$, $m_3 = \tan\theta_3$ and $m_4 = \tan\theta_4$; then $\Sigma m_1 = \frac{-2gf}{f^2 - c} = \Sigma \tan\theta_1 = s_1$ (say).

Similarly, $\Sigma \tan\theta_1 \tan\theta_2 = \frac{g^2 - 2ga - c}{f^2} = s_2$ (say).

$$\Sigma \tan\theta_1 \tan\theta_2 \tan\theta_3 = \frac{-2fa}{f^2 - c} = s_4 \text{ (say)}$$

Now as $\tan(\theta_1 + \theta_2 + \theta_3 + \theta_4) = \frac{s_1 + s_3}{1 - s_2 + s_4}$

$$= \frac{\frac{-2gf}{f^2 - c} + \frac{2fa}{f^2 - c}}{1 - \frac{g^2 - 2ag - c}{f^2 - c} - \frac{a^2}{f^2 - c}}$$

$$\Rightarrow \quad \tan(\Sigma\theta_1) = \frac{2f(g - a)}{(g - a)^2 - f^2}. \qquad ...(3)$$

Now as the co-ordinates of the centre and focus are respectively (g, f) and (a, 0), then

$$\tan \alpha = \frac{f}{g-a}. \qquad ...(4)$$

$$\text{Again, } \tan 2\alpha = \frac{2 \tan \alpha}{1-\tan^2 \alpha} = \frac{\frac{2f}{g-a}}{1-\left(\frac{2f}{g-a}\right)^2} = \frac{2f\,(g-a)}{(g-a)^2 - f^2} \qquad \text{[by (4)]}$$

So $\tan(\Sigma\theta_1) = \tan 2\alpha$.

Solving for most general value, we get

$\theta_1 + \theta_2 + \theta_3 + \theta_4 = n\pi + 2\alpha$, where n is an integer. **Proved.**

Example 105:

A parabola touches two given straight lines, which meet at O, in given points and a variable tangent meets the given lines in P and Q respectively; prove that the locus of the centre of the circumcircle of the triangle OPQ is a fixed straight line.

Solution:

Taking the given lines as axes, (x/f) + (y/g) = 1 as the equation of the variable tangent, we have (f/a) + (g/b) = 1.

Then the equation to OP will be x + y cos ω = f/2 and of the OQ, perpendicular to OP will be y + x cos ω = g/2.

The required locus will be the locus of the point of intersection of OP and OQ; so we have to eliminate g and f.

The required eliminant is $\frac{x + y\cos\omega}{a} + \frac{y + x\cos\omega}{b} = \frac{1}{2}$ which is a straight line. **Proved.**

Example 106:

Parabolas are drawn to touch two given rectangular axes and their foci are all at a constant distance c from the origin. Prove that the locus of the vertices of these parabolas is the curve $x^{2/3} + y^{2/3} = c^{2/3}$.

Solution:

We have $b = \lambda x^{1/3}$, $a = \lambda y^{1/3}$

$$\text{and} \qquad \lambda^2 = \frac{\left(x^{2/3} + y^{2/3}\right)^4}{x^{2/3}\, y^{2/3}} \qquad ...(1)$$

As co-ordinates of focus are $\left\{\frac{ab^2}{a^2+b^2}, \frac{a^2b}{a^2+b^2}\right\}$, square of the distance of focus from the origin (0, 0) is

$$\frac{a^2\,b^4}{\left(a^2+b^2\right)^2}+\frac{a^2\,b^2}{\left(a^2+b^2\right)^2}=\frac{a^2\,b^2}{\left(a^2+b^2\right)}=c^2 \quad \text{(by hypothesis)}$$

$$=\frac{\lambda x^{2/3}\,y^{2/3}}{x^{2/3}+y^{2/3}} \quad [\text{by (1)}] \qquad \ldots(2)$$

Eliminating λ from (1) and (2), we get $c^{2/3} = x^{2/3} + y^{2/3}$. **Proved.**

Example 107:

If a parabola, whose latus rectum is 4c, slide between two rectangular axes, prove that the locus of its focus is $x^2y^2 = c^2(x^2 + y^2)$ and that the curve traced out by its vertex is $x^{2/3}\,y^{2/3}\,(x^{2/3} + y^{2/3}) = c^2$.

Solution:

(a) As $\omega = 90°$, or $\cos\omega = 0$ so the focus is given by

$$x^2 + y^2 = ax = by$$

Also $\quad c^2\,(a^2 + b^2)^3 = a^4b^4.$

Eliminating a and b from (1) and (2), we get

$$c^2\left(x^2+y^2\right)^6\left(\frac{1}{x^2}+\frac{1}{y^2}\right)^3=\frac{\left(x^2+y^2\right)^8}{x^2\,y^2}$$

whence, simplifying, $\quad x^2y^2 = c^2\,(x^2 + y^2)$,

(b) The co-ordinates of vertex are

$$x=\frac{ab^4}{\left(a^2+b^2\right)^2},\; y=\frac{a^4b}{\left(a^2+b^2\right)^2} \qquad \ldots(3)$$

Hence $\quad \frac{x}{y}=\frac{b^3}{a^3}$ or $\frac{b}{x^{1/3}}=\frac{a}{y^{1/3}}=\lambda$ (suppose)

So $\quad a = \lambda y^{1/3}$ and $b = \lambda x^{1/3}$.

Putting the values of and b in (2),

$$c^2=\frac{\lambda^2\,x^{4/3}\,y^{4/3}}{x^{2/3}+y^{2/3}} \quad \text{and from (3),}$$

$$1=\frac{\lambda x^{1/3}\,y^{1/3}}{\left(x^{2/3}+y^{2/3}\right)^2}$$

Eliminating λ, we get $c^2 = x^{2/3}\,y^{2/3}\,(x^{2/3} + y^{2/3})$.

Example 108:

The axes being rectangular, prove that the locus of the focus of the parabola $\left(\frac{x}{a}+\frac{y}{b}-1\right)^2 = \frac{4xy}{ab}$, *a and b being variables such that a*

ab = c^2, *is the curve* $(x^2 + y^2)^2 = c^2xy$.

Solution:

The focus is given by $ax = by = x^2 + y^2$

and we have $ab = c^2$ (by hypothesis).

To eliminate a and b from (1) and (2), we have

$$\frac{x^2+y^2}{y} = b \text{ and } \frac{x^2+y^2}{x} = a$$

So $$ab = \frac{\left(x^2+y^2\right)^2}{xy} = c^2 \text{ by (2)}$$

$\Rightarrow$ $(x^2 + y^2)^2 = c^2xy$ which is the required locus.

Example 109(a):

Parabolas are drawn to touch the axes, which are inclined at an angle ω, and their directories all pass through a fixed point (h, k). Prove that all the parabolas touch the straight line $\frac{x}{h+k \sec \omega}+\frac{y}{k+h \sec \omega}=1$.

Solution:

Let the equation of the parabola be

$$\sqrt{(x/a)}+\sqrt{(y/b)}=1 \qquad ...(1)$$

The equation to the directories is

$$x\,(a + b \cos \omega) + y\,(b + a \cos \omega) = ab \cos \omega$$

Its (P's) distance from x-axis = y.

Its distance from origin *i.e.* (0, 0) = $\sqrt{[(x-0)^2+(y-0)^2}$

By hypothesis, $y=\frac{1}{2}\sqrt{\left(x^2+y^2\right)} \Rightarrow y=\frac{1}{4}\left(x^2+y^2\right)$ or $x^2 = 3y^2$.

As this line passes through (h, k), we have

$$h\,(a + b \cos \omega) + k\,(b + a \cos \omega) = ab \cos \omega \qquad ...(2)$$

The given line is $\frac{x}{h+k \cos \omega}+\frac{y}{k+h \cos \omega}=1$ ...(3)

If (1) touches (3), solving the two, we must get two coincident points.

By (1), $x = a/b\left(\sqrt{b}-\sqrt{y}\right)^2$

Putting in (3), we get $\frac{a\left(b+y-2\sqrt{by}\right)}{b\left(h+k\cos\omega\right)}+\frac{y}{k+h\cos\omega}=1$ ...(4)

or simplifying with the help of (2), we find (4) is a perfect square. Hence (3) touches (1).

Example 109(b):

Parabolas are drawn to touch two given straight lines which are inclined at an angle ω; if the chords of contact all pass through a fixed point, prove that (1) their directories all pass through another fixed point, and (2) their facial all lie on a circle which goes through the intersection of the two given straight lines.

Solution:

The equation of the directrix is written by

$$\frac{x+y\cos\omega}{a}+\frac{y+x\cos\omega}{b}=1 \qquad ...(1)$$

and as the chord of contact passes through a fixed point whose co-ordinates are (h, k), we obtain

$$\frac{h}{a}+\frac{k}{b}=1.$$

Therefore the directrix passes through the fixed point given by the following equation

$$x + y\cos\omega = h \text{ and } y + x\sec\omega = k.$$

And the focus is given by $ax = by = x^2 + y^2 + 2xy\cos\omega$, whence

$$a=\frac{x^2+y^2+2\,xy\cos\omega}{x} \quad \text{and} \quad b=\frac{x^2+y^2+2\,xy\cos\omega}{y}.$$

Substituting for 'a' and 'b' in (2) and simplifying, we have $x^2 + y^2 + 2xy\cos\omega - xh - yk = 0$, which is a circle which pass through the origin.

Example 109(c):

If the normals at the three points P, Q and R meet in a point and if PP', QQ' and RR' be chords parallel to QR, RP and PQ respectively, prove that the normals at P', Q', R' also meet in a point.

Solution:

Let P, Q and R be $\left(am_1^2,-2am_1\right)$, $\left(am_2^2,-2am_2\right)$ and $\left(am_3^2,-2am_3\right)$ respectively and P', Q' and R' be $\left(at_1^2,-2at_1\right)$, $\left(at_2^2,-2at_2\right)$ and $\left(at_3^2,2at_3\right)$ respectively.

Then Slope of $PP' = \frac{-2am_1 + 2at_1}{am_1^2 - at_1^2} = \frac{-2}{m_1 + t_1}$

and Slope of $QR = \frac{-2am_2 + 2am_3}{am_1^2 - am_3^2} = \frac{-2}{m_2 + m_3}$

As PP' as parallel to QR, slopes must be equal.

Hence $m_1 + t_1 = m_2 + m_3 = -m_1$ $\{\because \Sigma m_1 = 0\}$

$\therefore$ $t_1 = -2m_1$.

Similarly $t_2 = -2m_2$ and $t_3 = -2m_1 - 2m_2 - 2m_3$.

$= -2(m_1 + m_2 + m_3) = 0$ (as $\Sigma m_1 = 0$) or $\Sigma t_1 = 0$.

Hence the normals at point P', Q' and R' meet in a point.

Example 110:

A parabola is drawn such that each vertex of a given triangle is the pole of the opposite side; show that the focus of the parabola lies on the nine-point circle of the triangle and that the orthocentre of the triangle formed by joining the middle points of the sides lies on the directrix.

Solution:

The circle circumscribing the triangle formed by joining the middle points of the sides of the given triangle will be the nine point circle. As the circle circumscribing the triangle formed by the three tangents on the parabola always passes through the focus of the parabola (Ref. Art.179), hence, we have to prove that the lines joining the mid-points of the given triangle are the tangents on the parabola.

Let the equation of the parabola be $y^2 = 4ax$. ...(1)

Suppose the given triangle is ABC and let PQ be the chord of contact of A w.r.t. (1). Again, as BC is given to the polar of A w.r.t. (1), BC and PQ must lie on the same line. Hence the line joining the mid. points of AP and AQ also passes through the mid. points of AB & AC. Suppose the co-ordinates of P and Q be $(at_1^2, 2at_1)$ and $(at_2^2, 2at_2)$ respectively so that the co-ordinates of the point of intersection of the tangents at P and Q i.e. A are

$$[at_1t_2, a(t_1 + t_2)]$$

$\therefore$ Co-ordinates of mid-point of AP will be

$$\left\{\frac{at_1(t_1 + t_2)}{2}, \frac{a(3t_1 + t_2)}{2}\right\}$$

Similarly, the co-ordinates of mid, point of AQ will be

$$\left\{\frac{at_2(t_1 + t_2)}{2}, \frac{a(t_1 + 3t_2)}{2}\right\}$$

The line joining the mid-points of AP and AQ is

$$y-\frac{a(3t_1+t_2)}{2}=\frac{\frac{a(3t_1+t_2)}{2}-a\frac{(t_1+3t_2)}{2}}{\frac{at_1(t_1+t_2)}{2}-\frac{at_2(t_1+t_2)}{2}}\left\{x-\frac{at_1(t_1+t_2)}{2}\right\}$$

$$\Rightarrow \qquad y-\frac{2}{t_1+t_2}x+\frac{a(t_1+t_2)}{2}$$

This is a tangent on the parabola as it is of the form y = mx + a/m

Again the ortho-centre of the triangle formed by the tangents is on directrix (art. 181) and we have proved that the lines joining the mid. points of the sides of the triangle are tangents on the parabola.

Hence the ortho-centre of the triangle formed by joining the mid. Point of the sides of the triangle lies on directrix of the parabola.

Example 111:

A point on a parabola, the foot of the perpendicular from it upon the directrix, and the focus are the vertices of an-equilateral triangle. Prove that the focal distance of the point is equal to the latus rectum.

Solution:

Let the equation to the parabola be $y^2 = 4ax$ ...(1)

and P be any point on it as $(at^2, 2at)$. The focus S of the parabola will be (a, 0). If PN is the perpendicular from P on the directrix, x = – a; clearly the co-ordinates of N are (– a, 2at0 and the distance

$PN = (a + at^2)$ ...(2)

Distance $SN = \sqrt{[(-a-a)^2 + (2at-0)^2]}$

$= \sqrt{(4a^2 + 2a^2t^2)}$ ...(3)

If ΔPSN is an equilateral triangle, then

PN = NS [∵ PN = PS for all points]

Hence by (2) and (3), we get

$(a + at^2) = \sqrt{(4a^2 + 2a^2t^2)}$.

Squaring, we get $a^2 + a^2t^4 + 2a^2t^2 = 4a^2 + 4a^2t^2$

$\Rightarrow \quad t^4 - 2t^2 - 3 = 0$

$\Rightarrow \quad (t^2 - 3)(t^2 + 1) = 0.$

If $t^2 + 1 = 0$, t^2 is imaginary so, $t^2 - 3 = 0$ or $t = \pm\sqrt{3}$.

Hence the co-ordinates of P become $(3a, 2a\sqrt{3})$ or $(3a, -2a\sqrt{3})$.

Distance $PS = \sqrt{\{(3a-a)^2 + (2a\sqrt{3})^2\}} = \sqrt{(16a^2)} = 4a$

= latus rectum. **Hence proved.**

Example 112(a):

A series of chords is drawn so that their projections on a straight line which is inclined at an angle α to the axis are all of constant length c; prove that the locus of their middle point is the curve

$$(y^2 - 4ax)\,(y \cos \alpha + 2a \sin \alpha)^2 + a^2c^2 = 0.$$

Solution:

Let the equation of the parabola be $y^2 = 4ax$ ...(1)

and of the straight line having (x_1, y_1) as its mid-point be

$$\frac{x - x_1}{\cos\theta} = \frac{y - y_1}{\sin\theta} = r. \qquad ...(2)$$

By (2), we get $x = x_1 + r \cos \theta$ and $y = y_1 + r \sin \theta$.

Solving (1) and (2), we get

$$(y_1 + r \sin \theta)^2 = 4a\,(x_1 + r \cos \theta)$$

$$\Rightarrow \quad r^2 \sin^2 \theta + 2r\,(y_1 \sin \theta - 2a \cos \theta) + (y_1^2 - 4ax_1) = 0 \qquad ...(3)$$

(3) is quadratic in r.

The roots of this quadratic equation will be equal but of opposite sign ax (x_1, y_2) is the mid-point of the chord. So the coeff. of r must be zero.

Hence $2\,(y_1 \sin \theta - 2a \cos \theta) = 0$

$$\Rightarrow \quad \cot\theta = \frac{y_1}{2a} \qquad ...(4)$$

Again as the coeff. of r is zero, (iii) becomes

$$r^2 \sin^2 \theta = 4ax_1 - y_1^2 \quad \text{or} \quad r = \frac{\sqrt{(4ax_1 - y_1^2)}}{\sin\theta} \qquad ...(5)$$

Now length of the chord will be 2r. Let this chord and the given line make angle θ and α respectively with the axis.

Therefore, angle between the lines = θ – α, and the projection of the chord on the given line will be

$$2r \cos (\theta - \alpha) = c, \text{ (by hypothesis).}$$

$$\text{So} \qquad 2\sqrt{\left(\frac{4ax_1 - y_1^2}{\sin\theta}\right)} \cos (\theta - \alpha) = c, \qquad \text{[by (5)]}$$

$$\Rightarrow \quad 2\sqrt{(4ax_1 - y_1^2)}\ \frac{\cos\theta \cos\alpha + \sin\theta \sin a}{\sin\theta} = c$$

$$\Rightarrow \quad 2\sqrt{(4ax_1 - y_1^2)}\ (\cot \theta \cos \alpha + \sin \alpha) = c.$$

Putting the value of cot q from (4), we get

$$2\sqrt{(4ax_1 - y_1^2)}\left(\frac{y_1 \cos\alpha}{2a} + \sin\alpha\right) = c$$

$\Rightarrow$ $\sqrt{(4ax_1 - y_1^2)}$ (y. cos α + 2a sin α) = ac.

Squaring, we have

$$(4ax_1 - y_1^2)(y_1 \cos\alpha + 2a \sin\alpha)^2 = a^2c^2$$

$$\Rightarrow \quad (y_1^2 - 4ax_1)(y_1 \cos\alpha + 2a \sin\alpha)^2 = a^2c^2 = c.$$

Generalising this for (x_1, y_1), we get

$$(y^2 - 4ax)(y \cos\alpha + 2a \sin\alpha)^2 = a^2c^2 = c.$$ **Proved.**

Example 112(b):

Prove that all circles described on focal chords as diameters touch the directrix of the curve and that all circles on focal radii as diameters touch the tangent at the vertex.

Solution:

(a) Let the equation of the parabola be $y^2 = 4ax$.

The co-ordinates of the focus S will be (a, 0). Let PSQ be any focal chord, where the co-ordinates of P and Q are $(at_1^2, -2at1)$ and $(at_2^2, -2at_2)$ respectively.

Equation of PQ is $y(t_1 + t_2) = 2x + 2at_1t_2$. As it passes through (a, 0) the co-ordinates will satisfy it, hence

$$0 = 2a + 2at_1t_2 \quad \text{or} \quad t_1t_2 = -1 \quad \therefore \quad t_2 = -1/t_1.$$

So the co-ordinates of Q become $(a/t_1^2 - 2a/t_2)$.

Equation of circle drawn with PQ as diameter is

$$(x - at_1^2)(x - a/t_1^2) + (y - 2at_1)(y + 2a/t_1) = 0 \quad ...(1)$$

The equation of directrix is $x = -a$. ...(2)

Solving (1) and (2) simultaneously, we have

$$(-a - at_1^2)(-a - a/t_1^2) + (y - 2at_1)(y + 2a/t_1) = 0$$

$$\Rightarrow \quad y^2 - 2ay(t_1 \quad 1/t_1) + a^2(t_1 - 1/t_1)^2 = 0$$

$$\Rightarrow \quad \left\{y - a\left(t_1 - \frac{1}{t_1}\right)\right\}^2 = 0$$

As this is a perfect square, it will give us only one value of y. So (2) cuts (1) at two coincident points. Hence (2) is tangent to (1). **Proved.**

(b) Equation of circle drawn with line PS as diameter is

$(x - at_1^2)(x - a) + (y - 2at_1)(y - 0) = 0$

{as P is $(at_1^2, 2at_1)$ and S is $(a, 0)$}

$\Rightarrow \quad x^2 + y^2 - ax(1 + t_1^2) - 2at_1y + a^2t^2 = 0$ ...(3)

Equation to the tangent at vertex is $x = 0$.

Solving (3) and (4), we get

$$y^2 - 2ayt_1 + a^2t^2 = 0 \quad \text{or} \quad (y - at_1)^2 = 0$$

which is again a perfect square; hence the tow roots of y are coincident; therefore, the y-axis touches the circles, i.e. the tangent at the vertex touches (3).

Example 112(c):

A circle and a parabola intersect in four points; show that the algebraic sum of the ordinates of the four points is zero.

Show also that the line joining one pair of these pour points and the line joining the other pair are equally inclined to the axis.

Solution:

Let the equation to the circle be

$$x^2 + y^2 + 2gx + 2fy + c = 0 \quad \text{...(1)}$$

and of the parabola be $y^2 = 4ax$. ...(2)

On solving (1) and (2), we have [by (2) $x = y^2/4a$]

$$\frac{y^4}{16a^2} + y^2 + \frac{gy^2}{2a} + 2fy + c = 0 \quad \text{...(3)}$$

This is a fourth-degree equation in y. Let its roots be y_1, y_2, y_3 and y_4. Then sum of the roots

$$= -\frac{\text{coeff. of } y^3}{\text{coeff. of } y^4}.$$

Hence $\quad y_1 + y_2 + y_3 + y_4 = \dfrac{0}{1/16a^2} = 0$

Hence the sum of the ordinates of four points in which the two curves intersect is zero. Proved.

Again, let the points of intersection by P, Q, R and S having the co-ordinates as (x_1, y_1), (x_2, y_2), (x_3, y_3) and (x_4, y_4) resp. As all these points lie on the curve $y^2 = 4ax$, we have

Subtracting, we get $y_1^2 - y_2^2 = 4a(x_1 - x_2)$

$$\Rightarrow \quad \frac{y_1 - y_2}{x_1 - x_2} = \frac{4}{y_1 + y_2}. \quad \text{...(4)}$$

Now slope of the line PQ is say

$$m_1 = \frac{y_1 - y_2}{x_1 - x_2} = \frac{4a}{y_1 + y_2}. \qquad \text{[by (4)] ...(5)}$$

Similarly the slope of RS is say

$$m_2 = \frac{y_3 - y_4}{x_3 - x_4} = \frac{4a}{y_3 + y_4}. \qquad \text{...(6)}$$

As $y_1 + y_2 + y_3 + y_4 = 0$, so $y_3 + y_4 = -(y_1 + y_2)$

Putting in (6), $m_2 = \dfrac{4a}{-(y_1 + y_2)} = -m_1$

So m_1 and m_2 are equal in magnitude; hence pQ and RS are equally inclined to axis.

Example 113:

TP and TQ are tangents to the parabola and the normals are P and Q meet at a point R on the curve; prove that the centre of the circle circumscribing the triangle TPQ lies on the parabola

$$2y^2 = a(x - a).$$

Solution:

Let the parabola be $y^2 = 4ax$ and co-ordinates of P and Q be $(at_1^2, -2at_1)$ and $(at_2^2, -2at_2)$ respectively. At Q will be $t_2y = x + at_2^2$. Solving these two co-ordinates of the point of intersection T are $\{at_1t_2, a(t_1 + t_2)\}$. Now normal at P is $y = -t_1x + 2at_1 + at_1^3$.

Let this normal pass through R $(at_3^2, -2at_3)$ on the parabola. Then

$$t_1 + \frac{2}{t_1} = -t_3 \qquad \text{...(1)}$$

Similarly if normal at Q passes through $R, t_2 + \dfrac{2}{t_2} = -t_3$ ...(2)

So $\quad t_1 + \dfrac{2}{t_2} = t_2 + \dfrac{2}{t_2} = -t_3$

Subtracting, we get $(t_1 - t_2) + \dfrac{2}{t_1t_2}(t_2 - t_1) = 0$

$$\Rightarrow \quad (t_1 - t_2)\left(1 - \frac{2}{t_1t_2}\right) = 0$$

So either $t_1 = t_2$ which means P and Q coincide which is impossible

$$\Rightarrow \quad 1 - \frac{2}{t_1t_2} = 0$$

$$\Rightarrow \quad t_1t_2 = 2. \qquad \text{...(3)}$$

Again let the centre of the circle passing through P, Q and T be (x, y); we have

$$2x = a(t_1 + t_2)^2 + 2a \quad ...(4)$$

and $$2y = a(t_1 + t_2)(1 - t_1t_2) \quad ...(5)$$

Now, we have to eliminate $'t'_1$ and $'t'_2$ from these relations to find the locus of the centre of the circle.

By (5) and (3), we get

$$2y = a(t_1 + t_2)(1 - 2) = -a(t_1 + t_2); \quad \text{or} \quad t_1 + t_2 = \frac{-2y}{a}$$

Putting in (4), we get

$$2x = a\left(\frac{-2y}{a}\right)^2 + 2a \quad \text{or} \quad 2x = \frac{4y^2}{a} + 2a.$$

$$2y^2 = a(x - a).$$ **Proved.**

Example 114:

Prove that the locus of the middle points of all tangents drawn from points on the directrix to the parabola is

$$y^2 (2x + a) = a(3x + a)^2.$$

Solution:

Let P $(at^2, 2at)$ be any point on the parabola $y^2 = 4ax$, whose directrix is the line

$$x + a = 0. \quad ...(1)$$

Now tangent at P is $ty = x + at^2$. ...(2)

Solving (1) and (2), the co-ordinates of point of intersection are

$$\left(-a, \frac{at^2 - a}{t}\right)$$

Now, if (x, y) be the co-ordinates of the middle point of the tangent, then

$$2x = at^2 - a \quad ..(3)$$

and $$2y = \frac{at^2 - a}{t} + 2at$$

or $$2ty = 3at^2 - a.$$

Also $$6x = 3at^2 - 3a$$

Hence $$2ty = 6x + 2a$$

or $$ty = 3x + a \quad ...(4)$$

Eliminating t between (3) and (4), we have

$$(2x + a)y^2 = a(3x + a)^2$$

Example 115(a):

A circle is described on a focal chord as diameter; if m be the tangent of the inclination of the chord to the axis, prove that the equation to the circle is

$$x^2+y^2-2ax\left(1+\frac{1}{m^2}\right)-\frac{4ay}{m}-3a^2=0.$$

Solution:

Let the equation to the parabola be $y^2 = 4ax$.

If PSQ is any focal chord where S is the focus (a, 0) and P is any point $(at^2, 2at)$, the co-ordinates of Q will be $(a/t^2, -2a/t)$.

Now equation of the circle drawn with PQ as diameter is

$$(x - at^2)(x - a/t^2) + (y - 2at)(y + 2a/t) = 0$$

$$\Rightarrow \quad x^2 + y^2 - ax(t^2 + 1/t^2) - 2ay(t - 1/t) - 3a^2 = 0 \quad ...(1)$$

The slope of PQ = m (by hypothesis).

Hence $\quad m = \dfrac{2at+2a/t}{at^2-a/t^2} = \dfrac{2t}{t^2-1}$

$$\Rightarrow \quad 2/m = (t - 1/t). \quad ...(2)$$

Squaring, we get

$$\frac{4}{m^2} = t^2 + \frac{1}{t^2} - \quad \text{or} \quad t^2 + \frac{1}{t^2} = \left(\frac{4}{m^2} + 2\right).$$

Now eliminating t from equations (1) and (2), we have

$$x^2 + y^2 - ax(4/m^2 + 2) - 2ay(2/m) - 3a^2 = 0$$

$$\Rightarrow \quad x^2 + y^2 - 2ax(1 - 2/m^2) - (4ay/m) - 3a^2 = 0$$ **Proved.**

Example 115(b):

Circles are drawn through the vertex of the parabola to cut the parabola orthogonally at the either point of intersection. Prove that the locus of the centres of the circles is the curve

$$2y^2(2y^2 + x^2 - 12ax) = ax(3x - 4a)^2$$

Solution:

Let P $(at^2, 2at)$ be any point on the parabola $y^2 = 4ax$. The vertex A will be (0, 0). The equation of the tangent at p is

$$ty = x + at^2 \quad ...(1)$$

and the co-ordinates of the middle point of AP are $(at^2/2, at)$.

The slope of AP will be $\frac{2at-0}{at^2-0}=\frac{2}{t}$

The equation to the line through the point $\left(\frac{at^2}{2}, at\right)$

and perpendicular to AP will be $(y - at) = \frac{-t}{2}\left(x - \frac{at^2}{2}\right)$

$$\Rightarrow \quad t\left(x-\frac{at^2}{2}\right)+2(y-at)=0 \quad \text{or} \quad tx+2y=\frac{at^3}{2}+2at. \qquad ...(2)$$

Now the locus of the point of intersection of lines (1) and (2) will give us the locus of the centre of the which passes through the vertex A of the parabola.

from (1), $\frac{t^2y}{2}-\frac{tx}{2}=\frac{at^3}{2}$ and multiplying by $\frac{t}{2}$. ...(3)

Hence from (2) and (3), we have

$$t^2y + t(4a - 3x) - 4y = 0 \qquad ...(4)$$

Also (1) can be written as $t^2a - ty + x = 0$. ...(5)

Now we have to eliminate 'a' from (4) and (5) to get focus of the centre of the circle.

On cross-multiplication,

$$\frac{t^2}{x(4a-3y)-4y^2}=\frac{t}{-4ya-xy}=\frac{1}{-y^2-a(4a-3x)}$$

$$t=\frac{y(x+4a)}{y^2-a(3x-4a)} \text{ and } t^2=\frac{4y^2+x(3x-4a)}{y^2-a(3x-4a)}$$

hence $$\frac{y^2(x+4a)^2}{\left[y^2-a(3x-4a)\right]^2}=\frac{4y^2+x(3x-4a)}{y^2-a(3x-4a)}.$$

$$\Rightarrow \quad y^2(x+4y)^2=[\{4y^2+x(3x-4a)\}\{y^2-2(3x-4a)\}].$$

On simplifying, we get the locus of centre as

$$2y^2(2y^2+x^2-12ax)=ax(3x-4a)^2$$ **Proved.**

Example 115(c):

LOL' and MOM' are two chords of a parabola passing through a point O on its axis. Prove that the radical axis of the circles described on LL' and MM' as diameters passes through the vertex of the parabola.

Solution:

Let the equation of the parabola be $y^2 = 4ax$.

If O is any point on its axis, its co-ordinates may be taken as (b, 0). As LOL' and MCM' are two chords, suppose the co-ordinates or L be $(at_1^2, 2at_1)$, of L' be $(at_2^2, 2at_2)$, of M be $(at_3^2, 2at_3)$ and of M' be $(at_4^2, 2at_4)$.

The equation of LL' is $y(t_1 + t_2) = 2x + 2at_1t_2$; this passes through (b, 0); the co-ordinates will satisfy t; hence

$$0 = 2b + 2at_1t_2$$

or $\quad t_1t_2 = -b/a$

Similarly, $t_3t_4 = -b/a$.

So $\quad t_1t_2 = t_3t_4$

or $\quad t_1t_2 - t_3t_4 = 0.$ ...(1)

Now equation to the circles drawn with LL' as diameter is

$$(x - t_1^2)(x - at_2^2) + (y - 2at_1)(y - 2at_2) = 0 \quad \text{...(2)}$$

and the equation of the circle on MM' as diameter is

$$(x - t_3^2)(x - at_4^2) + (y - 2at_3)(y - 2at_4) = 0 \quad \text{...(3)}$$

Equation of radical axis of (3) and (2) is obtained by subtracting (3) from (2).

Subtracting (3) from (2), we get the constant term as

$$a^2(t_1^2t_2^2 + t_3^2t_4^2) + 4a^2(t_1t_2 + t_3t_4) = \lambda \text{ (say)}. \quad \text{...(4)}$$

By (1) and (4), we get $\lambda = 0$; hence the equation of the radical axis has no constant term; therefore this radical axis passes through the origin, the vertex of the parabolas. **Proved.**

Example 116:

The sides of a triangle touch a parabola and two of its angular points lie on another parabola with its axis in the same prove that the locus of the third angular point is another parabola.

Solution:

Suppose the points of contact of the side triangle on the parabola be $(at_1^2, 2at_1)$, $(at_2^2, 2at_2)$, and $(at_3^2, 2at_3)$.

Solving the tangents at these points, the co-ordinates of two are $\{at_2t_3,, a(t_2 + t_3)\}$ and $\{at_1t_3, a(t_1 + t_3)$.

Let these vertices lie on the other parabola $(y - h)^2 = 4b(x - 1)$.

Hence $\quad \{a(t_2 + t_3) - k\}^2 = 4b(at_2t_3 - h)$

and $\{a(t_3 + t_1) - k\}^2 = 4b(a_1t_3 - h)$.

Let the co-ordinates of third vertex be (x, y); we have

$$x = at_1t_2$$

and $y = a(t_1 + t_2)s$

To eliminate t_1, t_2 and t_3 from (1), (2), (3) and (4), we subtract from (1), we have $a(t_1 + t_2) + 2at_3 - 2k = 4bt_3$

Example 117:

Prove that the normal chord at the point whose ordinate is equal to its abscissa subtends a right angle at the focus.

Solution:

Let the equation of the parabola be $y^2 = 4ax$ ...(1)

and say P is any point on (1) for which the abscissa is equal to the ordinate is (h, h). Putting these co-ordinates in (1), we get $h^2 = 4ah$ hence either $h = 0$ or $h = 4a$.

Hence the point is either origin at which the normal is x-axis or (4a, 4a) at which the normal is

$$y - 4a = -\frac{4a}{2a}(x - 4a) \text{ or } y = -2x + 12a. \quad ...(2)$$

Putting the value of y in (1), we get

$$(-2x + 12a)^2 = 4ax$$

$$\Rightarrow \quad x^2 - 13ax + 36a^2 = 0 \text{ or } (x - 9a)(x - 4a) = 0$$

whence either $x = 9a$ or $4a$

The corresponding values of y are $-6a$ and $4a$.

Hence the other point of interaction of the normal at $P \equiv (4a, 4a)$ to the parabola (1) is (9a, 6a) say Q.

Slope of the line SP where S is the focus (a, 0) is

$$\frac{4a - 0}{2a - a} = \frac{4}{3} = m_1 \text{ (say)} \quad ...(3)$$

$$\text{slope of SQ} = \frac{-6a - 0}{+9a - a} = -\frac{3}{4} = m_2 \text{ (say)} \quad ...(4)$$

Clearly $m_1 \times m_2 = -1$, so the SP and SQ are at right angles.

Example 118:

The normal at the point $(at_1^2, 2at_1)$ meets the parabola again in the point $(at_2^2, 2at_2)$; prove that

$$t_2 = -t_1 - \frac{2}{t_1}.$$

Solution:

Let the equation to the parabola be $y^2 = 4ax$. ...(1)

The equation to the normal at any point $(at_1^2, 2at_1)$ to (1) will be

$$y - 2at_1 = \frac{-at_1}{2a}\left(x - at_1^2\right). \quad ...(2)$$

If (2) passes through the point $(at_2^2, 2at_2)$, the co-ordinates will satisfy it, hence

$$2at_2 - 2at_1 = -t_1\ (at_2^2, 2at_1^2)$$

$\Rightarrow$ $2a\ (t_2 - t_1) = -\ .\ a.\ (t_2 - t_1)\ (t_2 + t_1)$

$\Rightarrow$ $2 = -t_1 t_2 - t_1^2$

$\Rightarrow$ $t_1 t_2 = -2 - t_1^2$

$\Rightarrow$ $t_1 = -\frac{2}{t_1} - t_1.]$ **Proved.**

Example 119:

If PQ be a normal chord to the parabola and it S be the focus, prove that the locus of the centroid of the triangle SPQ is the curve

$$36y^2\ (3x - 5a) - 18y^4 = 128a^4.$$

Solution:

Let the equation to the parabola by $y^2 = 4ax$ ...(1)

and say P and Q are any two points $(at_1^2, 2at_1)$ and $(at_2^2, 2at_2)$ such that PQ is normal at P. Then by of this example

$$t_2 = -\frac{2}{t_1} - t_1. \quad ...(2)$$

Let the focus (a, 0) be S, as the centroid of any triangle having the vertices as (x_1, y_1), (x_2, y_2) and (x_3, y_3) is

$$\left(\frac{x_1 + x_2 + x_3}{3}, \frac{y_1 + y_2 + y_3}{3}\right)$$

so if the centroid of the triangle PQS will be (h, k) then

$$h = \frac{at_1^2 + at_2^2 + a}{3} \quad ...(3)$$

or $$k = \frac{at_1 + 2at_2 + 0}{3} \quad ...(4)$$

The required locus will be obtained by eliminating t_1 & t_2 from (2), (3) and (4).

So putting the value of t_2 from (2) in (4); we get

$$3k = 2at_1 + 2a\left(-t_1 - \frac{2}{t_1}\right) = -\frac{4a}{t_1} \quad \text{or} \quad t_1 = -\frac{4a}{3k}.$$

Hence $\quad t_2 = +\frac{4a}{3k} + \frac{2.3k}{4a}$

Putting the values of t_1 and t_2 in (2), we get

$$3h = a\left(\frac{4a}{3k}\right)^2 + a\left(\frac{4a}{3k} + \frac{2.3k}{4a}\right)^2 + a$$

$$\Rightarrow \quad 3h = \frac{16a^3}{9k^2} + \frac{16a^3}{9k^2} + \frac{9k^2}{4a} + 4a + a$$

$$\Rightarrow \quad 3h - 5a = \frac{32a^2}{9k^2} + \frac{9k^2}{4a}.$$

Simplifying further and generalising, we get

$$36ay^2(3x - 5a) = 128a^4 + 81y^4$$

$$36ay^2(3x - 5a) - 81y^4 = 128a^4$$

Example 120(a):

Prove that on the axis of any parabola there is a certain point K which has the property that, if a chord PQ of the parabola be drawn through it, then

$$\frac{1}{PK^2} + \frac{1}{QK^2}$$

is the same for all positions of the chord.

Solution:

Let the equation to the parabola be $y^2 = 4ax$. ...(1)

It axis is $y = 0$. So let there be any point K on the axis as (h, 0).

Equation of any line passing through (h, 0) is

$$y = mx - mh. \quad ...(2)$$

To get the point of intersection of (1) and (2) we solve them. So putting the value of y from (2) in (1), we get

$$(mx - mh)^2 = 4ax$$

$$\Rightarrow \quad m^2a^2 - 2x(m^2h + 2a) + m^2h^2 = 0$$

$$\Rightarrow \quad x = \frac{2\left(m^2h + 2a\right) \pm \sqrt{\left\{4\left(m^2h + 2a\right)^2 - 4.m^2.m^2h^2\right\}}}{2m^2}$$

$$= \frac{1}{m^2}\left\{\left(m^2 h + 2a\right) \pm \sqrt{\left(4a^2 + 4ahm^2\right)}\right\}.$$

The corresponding values of y are

$$\frac{1}{m}\left[2a \pm \sqrt{\left(4a^2 + 4ahm^2\right)}\right]$$

Hence the co-ordinates of P and Q are respectively

$$\left[\frac{\left(m^2 h + 2a\right) + \sqrt{\left(4a^2 + 4ahm^2\right)}}{m^2}, \frac{2a + \sqrt{\left(4a^2 + 4ahm^2\right)}}{m}\right]$$

and $\left[\dfrac{\left(m^2 h + 2a\right) - \sqrt{\left(4a^2 + 4ahm^2\right)}}{m^2}, \dfrac{2a - \sqrt{\left(4a^2 + 4ahm^2\right)}}{m}\right]$

So distance PK^2

$$= \left[\left\{\frac{\left(m^2 h + 2a\right) + \sqrt{\left(4a^2 + 4ahm^2\right)}}{m^2} - h + \frac{2a + \sqrt{\left(4a^2 + 4ahm^2\right)}}{m} - 0\right\}^2\right]$$

$$= \left[\left\{\frac{2a + \left(4a^2 + 4ahm^2\right)}{m^2}\right\}^2 + \left\{\frac{2a + \left(4a^2 + 4ahm^2\right)}{m}\right\}^2\right]$$

$$= \frac{\left\{2a + \sqrt{\left(4a^2 + 4ahm^2\right)}\right\}^2 \times \left(1 + m^2\right)}{m^4}$$

Similarly, $OK^2 = \dfrac{\left\{2a - \sqrt{\left(4a^2 + 4ahm^2\right)}\right\}^2 \times \left(1 + m^2\right)}{m^4}$

So $$\frac{1}{PK^2} + \frac{1}{Q^2} = \frac{m^2}{1 + m^2}\left[\left\{\frac{1}{2a + \sqrt{\left(4a^2 + 4ahm^2\right)}} +\right\}^2 + \left\{\frac{1}{2a - \sqrt{\left(4a^2 + 4ahm^2\right)}}\right\}^2\right]$$

$$= \frac{m^4}{1+m^2}\left[\frac{\left\{2a-\sqrt{\left(4a^2+4ahm^2\right)}\right\}^2+\left\{2a+\sqrt{\left(4a^2+4ahm^2\right)}\right\}^2}{\left\{4a^2-\left(4a^2+4ahm^2\right)\right\}^2}\right]$$

$$= \frac{m^4}{1+m^2}\left[\frac{4a^2+\_\ 4a^2+4ahm^2}{16a^2h^2m^2}\right] = \frac{2a+hm^2}{4ah^2\left(1+m^2\right)}.$$

If we put h = 2a, we get

$$\frac{1}{PK^2}+\frac{1}{QK^2}=\frac{2a+2am^2}{4a\,(2a)^2\left(1+m^2\right)}=\frac{2a\left(1+m^2\right)}{16a^3\left(1+m^2\right)}=\frac{1}{8a^2}$$

which is constant for all values of m. Hence for the point (2a, 0) the given property holds good.

Example 120(b):

Prove that the semi-latus-rectum is a harmonic mean between the segments of any focal chord.

Solution:

Let the equation of the parabola be $y^2 = 4ax$. ...(1)

The co-ordinates of focus are (a, 0). Let P be any point $(at_1^2, 2at_1)$ and Q be another point $(at_2^2, 2at_2)$, such that the line PQ passes through the focus S, then the equation to PQ will be

$$y - 2at_1 = \frac{2at_2 - 2at_1}{at_2^2 - at_1^2}\left(a - at_1^2\right)$$

$$\Rightarrow \qquad y - 2at_1 = \frac{2}{t_2+t_1}\left(a - at_1^2\right)$$

If this line passes through the focus (a, 0), we must have

$$0 - 2at_1 = \frac{2}{t_2+t_1}\left(a - at_1^2\right)$$

$$\Rightarrow \qquad t_1t_2 = -1 \qquad ...(2)$$

Distance SP = $\sqrt{[(at_1^2 - a)^2 + (2at_1)^2]} = (at_1^2 + a)$

Distance SQ = $\sqrt{[(at_2^2 - a)^2 + (2at_2)^2]} = (at_2^2 + a)$

The harmonic mean of SP and SQ will be

$$= \frac{2SP.SQ}{SP+SQ} = \frac{2\left(at_1^2+a\right)\left(at_2^2+a\right)}{\left(at_1^2+a\right)+\left(at_2^2+a\right)}$$

$$= \frac{2a^2\left(t_1^2+1\right)\left(\frac{1}{t_1^2}+1\right)}{a\left(2+t_1^2+\frac{1}{t_1^2}\right)} \qquad \left[\because = -\frac{1}{t_1} \text{ by } (2)\right]$$

$$= \frac{2a\frac{1}{t_1^2}\left(t_1^2+1\right)^2}{\frac{1}{t_1^2}\left(t_1^2+1\right)^2} = 2a = \text{semi-latus rectum.}$$

Example 120(c):

Prove that the normal chord at the point whose ordinate is equal to its abscissa subtends a right angle at the focus.

Solution:

Let the equation of the parabola be $y^2 = 4ax$ ...(1)

and say P is any point on (1) for which the abscissa is equal to the ordinate is (h, h). Putting these co-ordinates in (1), we get $h^2 = 4ah$ hence either $h = 0$ or $h = 4a$.

Hence the point is either origin at which the normal is x-axis or (4a, 4a) at which the normal is

$$y - 4a = -\frac{4a}{2a}(x-4a)$$

or $\qquad y = -2x + 12a.$ ...(2)

Putting the value of y in (1), we get

$$(-2x + 12a)^2 = 4ax$$

$\Rightarrow \qquad x^2 - 13ax + 36a^2 = 0$

or $\qquad (x - 9a)(x - 4a) = 0$

whence either $\qquad x - 9a$ or $4a$

The corresponding values of y are $-6a$ and $4a$.

Hence the other point of interaction of the normal at $P \equiv (4a, 4a)$ to the parabola (1) is (9a, 6a) say Q.

Slope of the line SP where S is the focus (a, 0) is

$$\frac{4a-0}{2a-a} = \frac{4}{3} = m_1 \text{ (say)} \qquad ...(3)$$

slope of SQ $= \dfrac{-6a-0}{+9a-a} = -\dfrac{3}{4} = m_2$ (say) ...(4)

Clearly $m_1 \times m_2 = -1$, so the SP and SQ are at right angles.

Example 121:

Prove that the area of the triangle formed by the normals to the parabola at the points $(at_1^2, 2at_1)$, $(at_2^2, 2at_2)$ and $(at_3^2, 2at_3)$ is

$$\frac{a^2}{2}(t_2-t_3)(t_3-t_1)(t_1-t_2)(t_1+t_2+t_3)^2.$$

Solution:

Let the equation of the parabola be $y^2 = 4ax$ and let these be any three points $(at_1^2, 2at_1)$, $(at_2^2, 2at_2)$ and $(at_3^2, 2at_3)$.

The equation of the normals at these points to the parabola are respectively

$$y = -t_1x + 2at_1 + at_1^2, \quad ...(1)$$

$$y = -t_2x + 2at_2 + at_2^2, \quad ...(2)$$

and

$$y = -t_3x + 2at_3 + at_3^2, \quad ...(3)$$

If (1) and (2) meet in A whose co-ordinates are (x_1, y_1), (2) and (3) in B whose co-ordinates are (x_2, y_2) and (3) and (1) in C whose co-ordinates are (x_3, y_3), then solving in pairs , we get

$$x_1 = 2a + a(t_1^2 + t_1t_2 + t_2^2),\ y_1 = -at_1t_2(t_1 + t_2),$$

$$x_2 = 2a + a(t_2^2 + t_2t_3 + t_3^2),\ y_2 = -at_2t_3(t_2 + t_3),$$

and

$$x_3 = 2a + a(t_3^2 + t_3t_2 + t_1^2),\ y_3 = -at_3t_1(t_3 + t_1).$$

Then area of the triangle ABC will be

$$\frac{1}{2}\begin{vmatrix} a(2+t_1^2+t_1t_2+t_2^2) & -at_1t_2(t_1+t_2) & 1 \\ a(2+t_2^2+t_2t_3+t_3^2) & -at_2t_3(t_2+t_3) & 1 \\ a(2+t_3^2+t_3t_1+t_1^2) & -at_3t_1(t_3+t_1) & 1 \end{vmatrix}$$

Subtracting Ist row from (2) and (3) and factorizing, we get

$$= -\frac{1}{2}a^2\begin{vmatrix} 2+t_1^2+t_1t_2+t_2^2 & t_1t_2(t_1+t_2) & 1 \\ (t_3-t_1)(t_3+t_2+t_1) & t_2(t_3-t_1)(t_1+t_2+t_3) & 0 \\ (t_3-t_2)(t_1+t_2+t_3) & t_1(t_3-t_2)(t_1+t_2+t_3) & 0 \end{vmatrix}$$

$$= -\frac{1}{2}a^2 \begin{vmatrix} (t_3 - t_1)(t_1 + t_2 + t_3) & t_3(t_2 - t_1)(t_1 + t_2 + t_3) \\ (t_3 - t_2)(t_1 + t_2 + t_3) & t_1(t_3 - t_2)(t_1 + t_2 + t_3) \end{vmatrix}$$

$$= -\frac{1}{2}a^2 (t_1 + t_2 + t_3)^2 (t_2 - t_1)(t_3 - t_2) \begin{vmatrix} 1 & t_2 \\ 1 & t_1 \end{vmatrix}$$

$$= \frac{1}{2}(t_1 - t_2)(t_2 - t_3)(t_3 - t_1)(t_1 + t_2 + t_3)^2$$ **Proved.**

Example 122:

The normal at a point P of a parabola meets the curve again in Q and T is the pole of PQ; show that T lies on the diameter passing through the other end of the focal chord passing through P, and that PT is bisected by the directrix.

Solution:

Let the equation of the parabola be $y^2 = 4ax$ and the co-ordinates of the point P be $(at_1^2, 2at_1)$ and of Q be $(at_2^2, 2at_2)$.

The equation of the line PQ the normal at P, will be

$$y = -t_1x + 2at_1 + at_1^3 \qquad ...(1)$$

If $T \equiv (\alpha, \beta)$ be the pole of PQ, then PQ will be the polar of T. Hence the equation of PQ must be

$$y\beta = 2a(x + \alpha). \qquad ...(2)$$

As (1) and (2) represent the same line PQ, so the coeffs. must be proportional, therefore,

$$\frac{\beta}{1} = \frac{2a}{-t_1} = \frac{2a\alpha}{2at_1 + at_1^2}. \qquad ...(3)$$

Let PR be the focal chord through P, where the co-ordinates of R are $(at_3^2, 2at_3)$. Then, the equation of PR will be

$$y - 2at_3 = \frac{2at_1 - 2at_3}{at_1^2 - at_3^2}\left(x - at_3^2\right)$$

or $$y - 2at_3 = \frac{2}{t_1 + t_3}\left(x - at_3^2\right)$$

As PR passes through the focus (a, 0), the co-ordinates will satisfy this equation.

Hence $$0 - 2at_3 = \frac{2}{t_1 + t_3}\left(a - at_3^2\right)$$

or $$t_3 = -\frac{1}{t_1}$$

So the co-ordinates of R become

$$\left(\frac{a}{t_1^2}, -\frac{2a}{t_1}\right)$$

The equation of the diameter through R will be

$$y = -\frac{2a}{t_1}. \quad ...(4)$$

By relation no (3),

$$\alpha = \left(-2a - at_1^2\right) \text{ and } \beta = -\frac{2a}{t_1}.$$

Hence the co-ordinates of T are $(-2a - at_1^2, -2a/t_1)$.

As these co-ordinates satisfy (4), T lies on (4) i.e. the diameter through R.

The equation of the directrix of the parabola

$$x + a = 0. \quad ...(5)$$

Abscissa of the mid-point of PT is

$$\frac{1}{2}\left(at_1^2 - 2a - at_1^2\right) = -a.$$

As the abscissa satisfies (5), the equation of the directrix, the line PT is bisected by the directrix.

Example 123:

If T be any point on the tangent at any point P of a parabola, and if TL be perpendicular to the focal radius SP and TN be perpendicular to the directrix, prove that SL = TN.

Hence obtain a geometrical construction for the pair of tangents drawn to the parabola from any point T.

Solution:

Let the equation to the parabola be $y^2 = 4ax$ ...(1)

and say $(at^2, 2at)$ is any point P on it. The equation to the tangent at P will be

$$y.2at = 2a\,(x + at^2)$$

or $$ty = x + at^2$$

If T is any point (x_1, y_1) on this line, then

$$ty_1 = x_1 + at^2. \quad ...(2)$$

Slope of the focal chord SP $= \dfrac{2at - 0}{at^2 - a} = \dfrac{2t}{t^2 - 1}.$

If TL is perpendicular to SP, then the equation of TL is

$$y - y_1 = -\frac{t^2 - 1}{2t}(x - x_1)$$

$$\Rightarrow \qquad 2ty - 2ty_1 + x(t^2 - 1) - x_1(t^2 - 1) = 0.$$

As SL is perp. to TL, the length of the perp.

$$SL = \frac{-2ty_1 + a(t^2 - 1) - x_1(t^2 - 1)}{\sqrt{\left[4t^2 + (t^2 - 1)^2\right]}} \qquad [\text{as} \equiv (a, 0)]$$

$$= \frac{-x_1 - at^2 - a - x_1t^2}{(t^2 + 1)} = \frac{(x_1 + a)(1 + t^2)}{(1 + t^2)}$$

Omitting the sign, we get SL = $(x_1 + a)$. ...(3)

If TN is the length of the perpendicular from T ≡ (x_1, y_1) on the directrix $x = -a$, then clearly TN = $(x_1 + a)$. ...(4)

by (3) and (4), we get SL = TN **Proved.**

Again

(i) If T is any point, from T draw perpendicular TN on the directrix AB.

(ii) Join TS where S is the focus.

(iii) Draw a circle on TS as diameter.

(iv) With S as centre, distance equal to TN, cut off two arcs from the circles say at L and M.

(v) Join SL and produce to meet the parabola at Q.

(vi) Join SM and produce to meet the parabola at Q.

(vii) Join TP and TQ.

(viii) TP and TQ are required tangents.

Example 124:

Parabola touches the sides of a triangle ABC in the points D. E and F respectively, prove that Bb and Cc are parallel.

Solution:

Let the parabola $y^2 = 4ax$ touch the sides BC, CA and AB of a triangle ABC at D, E and F respectively, and suppose the co-ordinates of the points D, E and F be $(at_1^2, 2at_1)$, $(at_2^2, 2at_2)$ and $(at_3^2, 2at_3)$ respectively.

So the tangents at D, E and F will represent the sides BC, CA and AB respectively.

Tangents at the points D, E and F are respectively

$$t_1y = x + at_1^2, \quad ...(1)$$

$$t_2y = x + at_2^2, \quad ...(2)$$

and $$t_3y = x + at_3^2, \quad ...(3)$$

On solving (2) and (3), (3) and (1) and (1) and (2) in pairs, we get the co-ordinates of A, B and C as $\{at_2t_3, a(t_2 + t_3)\};\{at_1t_3, a(t_1 + t_3)\}$ and $\{at_1t_2, a(t_1 + t_2)\}$ respectively. Now the diameter through the point A will be $y = a(t_1 + t_2)$ and equation of DE is $y(t_1 + t_2) = 2x + 2at_1t_2$ solving these equation, we have the co-ordinates of point of intersection b as $\{(a/2)\{(t_1^2 + t_2t_3 + t_1t_3 - t_1t_2), a(t_1 + t_2)\}$.

Again the equation of DF is $y(t_1 + t_3) = 2x + at_1t_3$; this cuts the diameter through A in c

By solving the two equation the co-ordinates of c are

$$\{(a/2)\{(t_2^2 + t_2t_3 + t_1t_2 - t_1t_3), a(t_2 + t_3)\}.$$

If slope of Bb m_1 and that of Cc be m_2, then

$$m_1 = \frac{2a(t_2 - t_1)}{a(t_2^2 + t_2t_3 - t_1t_3 - t_1t_2)} = \frac{2a(t_2 - t_1)}{a(t_1 + t_3)(t_2 - t_1)} = \frac{2}{(t_2 + t_3)}$$

Similarly slope of Cc is $m_2 = \dfrac{2}{(t_2 + t_3)}$

Since $m_1 = m_2$ hence the lines Cc and Bb are parallel. **Proved.**

Example 125(a):

T is the pole of the chord PQ; prove that the perpendicular from P, T and Q upon any tangent to the parabola are in geometrical progression.

Solution:

Let the equation of the parabola of $y^2 = 4ax$ and P and Q be any two points on it having the co-ordinates as $(at_2^2, 2at_2)$ and $(a_2^2, 2at_2)$ respectively; then the equation of the tangent at P will be and at Q will be

$$t_1y = x + at_1^2 \quad ...(1)$$

$$t_2y = x + at_2^2. \quad ...(2)$$

On solving (1) and (2); we get the co-ordinates of T as

$$(at_1t_2, a(t_1 + t_2)\}.$$

Any tangent of the parabola is $y = mx + (a/m)$. ...(3)

Length of the perpendicular TM from P on (3) is given by

$$PL = \frac{mat_1^2 - 2at_1 + a/m}{\sqrt{(1+m^2)}} = \frac{a(mt_1 - 1)^2}{m\sqrt{(1+m^2)}}$$

Length of the perpendicular TM from T on (3) is given by

$$TM = \frac{mat_1t_2 - a(t_1 + t_2) + a/m}{\sqrt{(1+m^2)}} = \frac{a(mt_1 - 1)(mt_2 - 1)}{m\sqrt{(1+m^2)}}$$

and perpendicular QN from Q on (3) (as in the case of PL) is given by

$$QN = \frac{a(mt_1 - 1)^2}{\sqrt{m(1+m^2)}}$$

$$\text{Now } LP \times QN = \frac{a^2(mt_2-1)^2(mt_1-1)^2}{m^2(1+m^2)} = TM^2$$

Hence PL, TM, CN are in GP **Proved.**

Example 125(b)

If from the vertex of a parabola a pair of chords be drawn at right angles to one another and with these chords as adjacent sides a rectangle be made, prove that the locus of the further angle of the rectangle is the parabola $y^2 = 4a(x - 8a)$.

Solution:

Let the equation of the parabola be $y^2 = ax$ and O be the vertex of it. If OP and OQ be the two adjacent sides of the rectangle OPRQ and the co-ordinates of P and Q be $(at_1^2, 2at_1)$ and $(at_2^2, 2at_2)$ resp; then the equation of the line PQ will be

$$y - 2at_1 = \frac{2at_2 - 2at_1}{at_2^2 - at_1^2}(x - at_1^2)$$

$$\Rightarrow \quad y(t_1 + t_2) = 2x + 2at_1t_2$$

$$\Rightarrow \quad \{y(t_1 + t_2) - 2x\}/2at_1t_2 = 1 \qquad ...(1)$$

The equation of the line joining the origin to the points of intersection of the parabola and (1) are obtained by making $y^2 = 4ax$ homogenous with the help of (1); so, we get

$$y^2 = 4a.x.\frac{y(t_1 + t_2) - 2x}{2at_1t_2}$$

$$2at_1t_2y^2 - 4a(t_1 + t_2)xy + 8ax^2 = 0.$$

If the lines are at right angles, then the sum of the coeff. of x^2 and y^2 should be zero, hence

$$2at_1t_2 + 8a = 0$$

or $\qquad t_1t_2 + 4 = 0$ ...(2)

If M is the mid-point of PR, then its co-ordinates will be

$$\left(\frac{at_1^2+at_3^2}{2}, \frac{2at_1+2at_3}{2}\right)$$

Let the co-ordinates of the reqd, point R be (h, k) then M will be the mid-point of OA; also as CPRQ is a rectangle, so, we get

$$\frac{h+0}{2}=\frac{at_1^2+at_2^2}{2} \quad \text{or} \quad h = a\left(t_1^2+t_2^2\right) \qquad ...(3)$$

and
$$\frac{k+0}{2}=\frac{at_1+2at_2}{2} \quad \text{or} \quad k = 2a\left(t_1+t_2\right) \qquad ...(4)$$

To get the reqd. locus, we have to eliminate t_1 and t_2 from (2), (3) and (4), so squaring (4), we get

$$k^2 = 4a^2 (t_1^2 + t_2^2 + 2t_1t_2)$$

$$= 4a^2\left\{\frac{h}{a}+2.(-4)\right\},$$

{as $t_1^2 + t_2^2 = h/a$ by (1)

and $\quad 2t_1t_2 = -4$ by (2)} $\Rightarrow k^2 = 4ah - 32a^2$

Generalising, we get the reqd. locus as

$$y^2 = 4a\ (y - 8a).$$ **Proved.**

Example 125(c):

Prove that the locus of the poles of chords which subtend a right angle at a fixed point (h, k) is

$$ax^2 - hy^2 + (4a^2 + 2ah)\,x - 2aky + a\,(h^2 + k^2) = 0.$$

Solution:

Let $(at_1^2, 2at_1)$ and $(at_2^2, 2at_2)$ be the ends of the chord PQ, with respect to parabola $y^2 = 4ax$, and (x_1, y_1) be its, pole.

Equation of PQ will be $y - 2at_1$

$$= \frac{2at_2-2at_1}{at_2^2-at_1^2}\left(x-at_1^2\right)$$

$\Rightarrow \qquad y\,(t_1 + t_2) = 2ax + at_1t_2$ ...(1)

Again polar of (x_1, y_1) w.r.t. the parabola $y^2 = 4ax$ is

$$yy_1 = 2a\,(x + x_1) \qquad ...(2)$$

As (1) and (2) represent the same straight line, their coeffs. will be proportional. Hence comparing the coeffs.

$$\frac{y_1}{t_1+t_2} = \frac{2a}{2} = \frac{2ax_1}{2at_1t_2}$$

whence $x_1 = at_1t_2$ ...(3)

and $y_1 = a\ (t_1 + t_2)$. ...(4)

Let (h, k) be any point A. The slopes of PA and PB are respectively

$$\frac{k-2at_1}{h-at_1^2} \text{ and } \frac{k-2at_2}{h-at_2^2}$$

If PAQ = 90°, we have $\dfrac{k-2at_1}{h-at_1^2} \times \dfrac{k-2at_2}{h-at_2^2} = -1$

or $k^2 - 2ak\ (t_1 + t_2) + 4a^2t_1t_2 + h^2 - a\ (t_1^2 + t_2^2)\ h + a\ a^2t_1^2t_2^2 = 0$ ...(5)

From (3) and (4), $a^2\ (t_1^2 + t_2^2) = y_1^2 - 2ax_1$. ...(6)

Eliminating t_1 and t_2 from (3), (4), (5) and (6), we have

$$k^2 - 2ky_1 + 4ax_1 + h^2 - h\left(\frac{y_1^2-2ax_1}{a}\right) + x_1^2 = 0$$

Generalising and simplifying, the locus of the pole (x_1, y_1) is given by

$$ax^2 + hy^2 + (4a^2 + 2ah)\ x - 2aky + a\ (h^2 + k^2) = 0.$$

Example 125(d):

If r_1 and r_2 be the lengths of radii vectors of the parabola which are drawn at right angles to one another from the vertex, prove that

$$r_1^{4/3}\ r_2^{4/3} = 16a^2\ (r_1^{2/3} + r_2^{2/3}).$$

Solution:

Let the parabola be $y^2 = 4ax$ ...(1)and let P and Q be any two points on it such that the co-ordinates of P and Q are $(r_1 \cos\theta, r_1 \sin\theta)$ and $(r_2 \sin\theta, r_2 \cos\theta)$ respectively.

$(\because \angle PAQ = 90°)$.

Since the points lie on (1), the co-ordinates will satisfy it.

So $r_1^2 \sin^2\theta = 4\ ar_1 \cos\theta$

$\Rightarrow$ $r_1 = 4a \cos\theta/\sin^2\theta$

and $r_1^2 \cos^2\theta = 4ar_2 \sin\theta$

$\Rightarrow$ $r_2 = 4a \sin\theta/\cos^2\theta$.

Now $$r_1r_2 = \frac{16a^2 \cos\theta \sin\theta}{\sin^2\theta \cos^2\theta} = \frac{16a^2}{\sin\theta \cos\theta}$$

$$\Rightarrow \qquad r_1^{2/3}\, r_2^{4/3} = \frac{\left(16a^2\right)^{4/3}}{\sin^{4/3}\theta\,\cos^{4/3}\theta} \qquad ...(2)$$

and $$r_1^{2/3}\, r_2^{2/3} = (4a)^{2/3}\left\{\frac{\cos^{2/3}\theta}{\sin^{4/3}\theta} + \frac{\sin^{2/3}\theta}{\cos^{4/3}\theta}\right\}$$

or $$r_1^{2/3}\, r_2^{2/3} = \frac{(4a)^{2/3}}{\sin^{4/3}\theta\,\cos^{4/3}\theta}\left(\sin^2\theta + \cos^2\theta\right)$$

$$= \frac{\left(16a^2\right)^{1/3}}{\sin^{4/3}\theta\,\cos^{4/3}\theta} \qquad ...(3)$$

Now dividing (2) by (3), we have

$$\frac{r_1^{4/3}\, r_2^{4/3}}{\left(r_1^{2/3} + r_2^{2/3}\right)} = 16a^2;\ \text{ or }\ r_1^{4/3}\, r_2^{4/3} = 16a^2\left(r_1^{2/3} = r_2^{2/3}\right)$$ **Proved.**

Example 126:

Prove that the equation to the circle passing through the points $(at_1^2, 2at_1)$ and $(at_2^2, 2at_2)$ and the intersection of the tangents to the parabola at these points is

$$x^2 + y^2 - ax\,[(t_1 + t_2)^2 + 2] - ay\,(t_1 + t_2)\,(1 - t_1t_2) + a^2t_1t_2\,(2 - t_1t_2) = 0.$$

Solution:

Let $y^2 = 4ax$ be the parabola on which the points P $(at_1^2, 2at_1)$ and Q $(at_2^2, 2at_2)$ lie. The equation of the tangent at P will be

$$t_1y = x + at_1^2. \qquad ...(1)$$

and at Q will be $$t_2y = x + at_2^2. \qquad ...(2)$$

Solving (1) and (2), we get the co-ordinates of T, the point of intersection of tangents as $[at_1t_2,\ a\,(t_1 + t_2)$. Now all these three points lie on the circle.

Let the equation of the circle be

$$x^2 + y^2 + 2gx + 2fy + c = 0. \qquad ...(3)$$

As T, P and Q lie on (3), the co-ordinates will satisfy it, hence taking one by one, we get

$$a^2t_1^2t_2^2 + a^2\,(t_1 + t_2)^2 + 2agt_1t_2 + 2af\,(t_1 + t_2) + c = 0, \qquad ...(4)$$

$$a^2t_1^4 + 4a^2t_1^2 + 2agt_1^2 + 4aft_1 + c = 0 \qquad ...(5)$$

$$a^2t_2^4 + 4a^2t_2^2 + 2agt_2^2 + 4aft_2 + c = 0 \qquad ...(6)$$

Now we have to solve these 3 equations.

From the sum of (5) and (6), subtracting twice equation (1), we get

$$a\,(t_1 + t_2)^2 + 2a + 2g = 0$$

or $$2g = -\,[\,(t_1 + t_2)^2 + 2]$$

Again subtracting (6) from (5); we get

$$a\,(t_1 + t_2)\,(t_1^2 + t_2^2) + 4a^2\,(t_1 + t_2) + 2g\,(t_1 + t_2)^2 = -\;4f.$$

Substituting for g, we have $2f = -\,(t_1 + t_2)\,(1 - t_1t_2)$.

Substituting for 'g' and 'f' in (5), we have

$$a^2t_1^4 + 4a^2t_1^2 \;-\; a^2t_1t_2(t_1^2 + t_2^2 + 2t_1t_2 - 2a^2t_1^2 - 2a^2t_1^2$$
$$-\,(1 - t_1t_2) \times (t_1 + t_2) + c = 0$$

Whence $c = a^2t_1t_2\,(2 - t_1t_2)$

Putting these values in (3), we get the equation to the circle as

$$x^2 + y^2 - ax\,(t_1 + t_2)^2 + 2] - ay\,(t_1 + t_2)\,(1 - t_1t_2)$$
$$+\,a^2t_1t_2\,(2 - t_1t_2) = 0.$$

Example 127:

The general equation to a system of parallel chords in the parabola $y^2 = (25/7)\,x$ is $4x - y + k = 0$. what is the equation to the corresponding diameter

Solution:

The equation of the parabola is given as

$$y^2 = \frac{25}{7}x \qquad ...(1)$$

and a line is given as $4x - y + k = 0.$ ...(2)

To solve (1) and (2), we put the value of x from (2) in (1); we get

$$y^2 = \frac{25}{7}\,\frac{y-k}{4}$$

$$\Rightarrow \qquad 28y^2 - 25y + 25k = 0.$$

If the ordinates of the point of intersection of (1) and (2) be y_1 and y_2 respectively; then y_1 and y_2 are the roots of (3).

So $$y_1 + y_2 = \frac{25}{28}$$

If the ordinate of the mid-point of the line joining the points whose ordinates are y_1 and y_2 is

$$\frac{y_1 + y_2}{2} = \frac{25}{28 \times 2} = \frac{25}{56}.$$

As the diameter of (1) corresponding to (2) is a line parallel to the axis of (1) i.e. y = 0 (x-axis) through any chord parallel to (2) will be passing through the point whose ordinate is 25/56 so its equation is y = 25/56 or 56y = 25.

Example 128:

If ω be the angle which a focal of a parabola makes with the axis, prove that the length of the chord is 4a cosec² ω and that the perpendicular on it from the vertex is a sin ω.

Solution:

Let the equation of the parabola by $y^2 = 4ax$. ...(1)

Focus will be (a, 0) and any line passing through (a, 0) may be given by

$$y = m(x - a) \quad ...(2)$$

To get the points of intersection of (1) and (2), we solve them, so putting the value of x from (2) in (1), we get

$$y^2 = 4a\frac{y + am}{m}$$

or $my^2 - 4ay - 4am = 0.$ ...(3)

If y_1 and y_2 are the roots of (3), then

$$y_1 + y_2 = 4a/m \quad ...(4)$$

and $y_1 + y_2 = -4a.$ ...(5)

Let (2) intersect (1) in A and B, whose co-ordinates are (x_1, y_1) and (x_2, y_2); then

$$AB = \sqrt{[(x_2 - x_1)^2 + (y_2 - y_1)^2]} \quad ...(6)$$

As (x_1, y_1) and (x_2, y_2) lie on (2), so

$$y_1 = m(x_1 - a) \quad ...(7)$$

$$y_2 = m(x_2 - a) \quad ...(8)$$

Subtracting (7) from (8), we get

$$(y_2 - y_1) = m(x_2 - x_1)$$

or $$(x_2 - x_1) = \frac{1}{m}(y_2 - y_1).$$

Putting this will in (4), we get

$$AB = \sqrt{\left[\frac{1}{m}(y_2 - y_1)^2 + (y_2 - y_1)^2\right]} = \sqrt{\left[(y_2 - y_1)^2\left(\frac{1}{m} + 1\right)\right]}$$

$$= \sqrt{[\{(y_2 + y_1)^2 - 4y_1y_2\}\ \{(\cot^2 \omega + 1)\}]}$$

[as the angle that (2) makes with x-axis is given to be ω so m = tan ω]

$$=\sqrt{\left[\left\{\left(\frac{4a}{\tan\omega}\right)^2-(-16a^2)\right\}\operatorname{cosec}^2\omega\right]} \quad \text{[by (4) and (5)]}$$

$= \sqrt{[16a^2 (\cot^2 \omega + 1) \operatorname{cosec}^2 \omega]}$

$= \sqrt{[16a^2 \operatorname{cosec}^2 \omega \operatorname{cosec}^2 \omega]} = 4a \operatorname{cosec}^2 \omega$

Again the length perpendicular from vertex (0, 0) on (2)

$$=\frac{am}{\sqrt{(1+m^2)}}=\frac{a\tan\omega}{\sqrt{(1+\tan^2\omega)}}=\frac{a\sin\omega}{\cos\omega.\sec\omega}=a\sin\omega.$$

Example 129(a):

Find the locus of a point O when the three normals drawn from it are such that two of them make complementary angles with the axis.

Solution:

Let the equation to the parabola by $y^2 = 4ax$; then the equation to any normal to it will be $y = mx - 2am - am^3$.

If this normal passes through (h, k), the co-ordinates will satisfy, hence $k = mh - 2am - 2am - am^3$ or $am^3 + (2a - h) m + k = 0$.

Let m_1, m_2 and m_3 be the roots of this equation, then

$$m_1 + m_2 + m_3 = 0, \quad ...(1)$$

$$m_1 m_2 + m_2 m_3 + m_3 m_1 = \frac{2a-h}{a}$$

and $$m_1 m_2 m_3 = -\frac{k}{a}. \quad ...(3)$$

Let one of the normals make an angle θ with axis of x, then the other will make an angle (90° – θ) by hypothesis.

So $m_1 = \tan\theta$, and $m_2 \tan(90 - \theta) = \cot\theta$.

So $m_1 m_2 = 1$. ...(4)

Now we have to eliminate m_1, m_2 and m_3 from (1), (2), (3) and (4).

Now from (3) and (4), $m_3 = -\frac{k}{a}$. ...(5)

By (2), $m_1 m_2 + m_3 (m_1 + m_2) = \frac{2a-h}{a}$

or $m_1 m_2 + m_3 (-m_3) = \frac{2a-h}{a}$ [by (1), $m_1 + m_2 + m_3 = 0$] ...(6)

Putting the value of m_1m_2 from (4) and m_2 from (5) we have from (1),

$$1-\frac{k^2}{a^2}=\frac{2a-h}{a} \text{ or } k^2 = a\ (h - a).$$

Generalising, the locus of (h, k) is $y^2 = a\ (x - a)$. **Ans.**

Example 129(b):

If OP and OQ make complementary angles withe axis, then the tangent of R is parallel to SO.

Solution:

Let the equation of any normal OP be $y = m_1x - 2am_1 - am_1^3$, and the equation of the other normal OQ be $y = m_2x - 2am_2 - am_2^3$. Let OP be inclined at an angle of θ and OQ at an angle of ϕ. Then θ + ϕ = 90° or ϕ = (90° − θ) and if $m_1 = \tan\theta$ and $m_2 = \tan(90° - \theta) = \cot\theta$.

$\therefore \quad m_1m_2 = 1.$ ...(1)

The co-ordinates of the focus S of the parabola will be (a, 0) and those of O are (h, k) (given).

Hence the slope of SO $= \dfrac{k-0}{h-a} = \dfrac{k}{h-a}$. ...(2)

The normal at R is $y = m_3x - 2am_3 - am_3^3$. Hence the slope of the tangent at R is $= -1/m_3$.

Again, $m_1 + m_2 + m_3 = 0$, $m_1m_2 + m_2m_3 + m_1m_3 = \dfrac{2a-h}{a}$

and $m_1\ m_2\ m_3 = -\dfrac{k}{a}$.

Hence $m_3(m_1 + m_2) + m_1\ m_2 = \dfrac{2a-h}{a}$

or $-m_3^2+1=\dfrac{2a-h}{a} \quad \therefore \quad -m_3^2=\dfrac{a-h}{a}$. ...(3)

Again $m_3 = \dfrac{-k}{am_1\ m_2} = \dfrac{-k}{a}\left[\text{as } m_1\ m_2 = 1 \text{ by (1)}\right]$. ...(4)

Dividing equation (4) by equation (3), we get $\dfrac{-1}{m_3}=\dfrac{k}{a-1}$.

Hence the slope of the tangent at R is same as the slope of SO.

∴ SO is parallel to the tangent at R.

Example 129(c):

Two equal parabolas have the same focus and their axes are at right angles; a normal to one is perpendicular to a normal to the other; prove that the locus of the point of intersection of these normals is another parabola.

Solution:

Taking the common focus as origin and axis of the parabolas as axis of the co-ordinates, the equations of the parabolas may be written as

$$y^2 = 4a\,(x + a) \qquad ...(1)$$

and $$x^2 = 4a\,(y + a) \qquad ...(2)$$

where 4a is the latus rectum of each.

Equation to any normal to (1) is

$$y = m\,(x + a) - 2am - am^3. \qquad ...(3)$$

Equation to any normal to (2) is

$$y = m'\,(y + a) - 2am' - am'^3. \qquad ...(4)$$

But the two normal are at right angles, then

$$m.\frac{1}{m} = -1 \quad \text{or} \quad m' = -m \left[\text{as the slope of (4) is } \frac{1}{m'}\right]$$

Substituting m' = – m in (4), we have

$$x = -m\,(y + a) + 2\,2am + am^3 \qquad ...(5)$$

The locus of the point of intersection of the normals will be the eliminant of m from (3) and (5). So adding (3) and (5), after taking the terms to R.H.S. in both the cases, we get

$$0 = mx - x - y - my$$

$$\Rightarrow \quad x + y = m\,(x - y)$$

or $$m = \frac{x+y}{x-y}.$$

Putting the value of m in (3), we get

$$y = \frac{x+y}{x-y}.(x+a) - 2a.\frac{x+y}{x-y} - a\left(\frac{x+y}{x-y}\right)^3$$

$$\Rightarrow y\,(x - y)^3 = (x + y)\,[(x + a)\,(x - y)^2 - 2a\,(x - y)^2 - a\,(x + y)^2]$$

$$= (x + y)\,[(x - y)^2\,(x + a - 2a) - a\,(x + y)^2]$$

$$= (x + y)\,[(x\,(x - y)^2 - 2a\,(x^2 + y^2)]$$

$$\Rightarrow 2a\,(x^2 + y^2)\,(x + y) = (x + y).\,x\,(x - y)^2 - y\,(x - y)^3$$

$$\Rightarrow 2a\,(x^2 + y^2)\,(x + y) = (x - y)^2\,[x^2 + xy - xy + y^2]$$

$$\Rightarrow 2a\,(x + y) = (x - y)^2.$$

This is the reqd. locus which is a parabola. **Proved.**

Example 130:

If $a^2 > 8b^2$, prove that a point can be found such that the two tangents from it to the parabola $y^2 = 4ax$ are normals to the parabola $x^2 = 4by$.

Solution:

Two parabolas are given as

$$y^2 = 4ax \qquad ...(1)$$

and $$x^2 = 4by \qquad ...(2)$$

Tangent to (1) $y = mx + \dfrac{a}{m}$. ...(3)

Equation to any normal sto (2) is

$$x = m'y - 2bm' - bm'^3$$

$$\Rightarrow \quad y = \frac{1}{m'}x + 2b + bm'^2 \qquad ...(4)$$

If the equations (3) and (4) are the same, then the coefficients must be identical. As the coefficients of y are equal in both, so coefficients of x and constant, terms will also be equal. Hence,

$$m = \frac{1}{m'} \quad \text{or} \quad m' = \frac{1}{m} \qquad ...(5)$$

and $$\frac{a}{m} = 2b + bm'^2. \qquad ...(6)$$

Putting the value of m' from (5) in (6), we get

$$\frac{a}{m} = 2b + \frac{b}{m^2} \qquad \text{or} \qquad 2bm^2 - am + b = 0. \qquad ...(7)$$

If the roots of (7) are real, then its discriminate must be greater then zero.

Therefore $a^2 - 4.\ 2b.b \geq 0$ or $a^2 \geq 8b^2$,

which is the given condition.

5

Convergence of Improper Integrals

INTRODUCTION

When the limit of an improper integral as defined above, is a definite finite number, we say that the given integral is *convergent* and the value of the integral is equal to the value of that limit. When the limit is ∞ or $-\infty$, the integral is said to be *divergent i.e.*, the value of the integral does not exist.

In case the limit is neither a definite *number* nor ∞ or $-\infty$, the integral is said to be *oscillatory* and in this case also the value of the integral does not exist *i.e.*, the integral is not convergent. We can define the convergence of the infinite integrand $\int_a^\infty f(x)dx$ as follows:

Definition: *The integral* $\int_a^\infty f(x)dx$ *is said to converge to the value I, if for any arbitrarily chosen positive number* ε*, however small but not zero, there exists a corresponding positive number N such that*

$$\left|\int_a^b f(x)dx - I\right| < \varepsilon \text{ for all values of } b \geq N.$$

Similarly, we can define the convergence of an integral, when the lower limit is infinite, or when the integrand becomes infinite at the upper or lower limit.

TESTS FOR CONVERGENCE OF IMPROPER INTEGRALS OF THE FIRST KIND

i.e, to test the convergence of improper integrals in which the range of integration is infinite and the integrand is bounded.

If an integral of the form $\int_a^\infty f(x)dx$ or $\int_{-\infty}^b f(x)dx$ cannot be actually integrated, its convergence is determined with the help of the following tests:

THE M-TEST

Let f (x) be bounded and integrable in the interval (a, ∞) where a > 0.

If there is a number m > 1, such that $\lim_{x \to \infty} x\mu\, f(x)$ *exists, then* $\int_a^{\infty} f(x)dx$ *is convergent.*

If there is a number $\mu \leq 1$*, such that* $\lim_{x \to \infty} x\mu\, f(x)$ *exists and is non-zero, then* $\int_a^{\infty} f(x)dx$ *is divergent and the same is true if* $\lim_{x \to \infty} x\mu\, f(x)$ *is* $+\infty$ *or* $-\infty$.

While applying the μ-test, the value of μ is usually taken to be equal to the highest power of x in the denominator of the integrand minus the highest power of x in the numerator of the integrand.

THE μ-TEST

Let f (x) be unbounded at x = a and be bounded and integrable in the arbitrary interval

(a + ε, b), where 0 < ε < b − a.

If there is a number μ between 0 and 1 such that

$$\lim_{x \to a+0} (x-a)^{\mu} f(x) \text{ exists, then } \int_a^b f(x)dx \text{ is convergent.}$$

If there is a number $\mu \geq 1$ *such that* $\lim_{x \to a+0} (x-a)^{\mu} f(x)$ *exist and is non-zero, then* $\int_a^b f(x)dx$ *is divergent and the same is true if* $\lim_{x \to a+0} (x-a)\mu\, f(x) = +\infty$ *or* $-\infty$.

In case f (x) is unbounded at x = b, we should find

$$\lim_{x \to b+0} (b-x)^{\mu}.f(x),$$

the other conditions of the test remaining the same.

COMPARISON TEST

Let f (x) and g (x) be two functions which are bounded and integrable in the interval (a, ∞). Also let g (x) be positive and $|f(x)| \leq g(x)$ *when* $x_0 > a$*. Then, if* $\int_a^{\infty} g(x)dx$ *is convergent,* $\int_a^{\infty} f(x)dx$ *is also convergent.*

Similarly if $|f(x)| \geq g(x)$ *for all values of x greater than some number x0 < a and* $\int_a^{\infty} g(x)dx$ *is divergent, then* $\int_a^{\infty} g(x)dx$ *is also divergent.*

Alternative Form of the Above Comparison Test

If $\lim_{x \to \infty} \frac{f(x)}{g(x)}$ *is a definite number, other than zero, the integrals* $\int_a^{\infty} f(x)dx$ *and* $\int_a^{\infty} g(x)dx$ *either both converge of both diverge.*

Note: While applying comparison test, we generally take $g(x) = \frac{1}{x^n}$ *i.e.,* $\int_a^{\infty} \frac{dx}{x^n}$ is generally taken as the comparison integral.

Theorem:

The comparison integral $\int_a^{\infty} \frac{dx}{x^n}$*, where a > 0, is convergent when > 1 and divergent when* $n \leq 1$.

Proof:

By the definition of an improper integral, we have

$$\int_a^{\infty} \frac{dx}{x^n} = \lim_{X \to \infty} \int_a^{X} \frac{dx}{x^n} = \lim_{X \to \infty} \int_a^{X} x^{-n} dx$$

$$= \lim_{X \to \infty} \left[\frac{x^{1-n}}{1-n} \right]_a^X, \text{ if } n \neq 1$$

$$= \lim_{X \to \infty} \left[\frac{X^{1-n}}{1-n} - \frac{a^{1-n}}{1-n} \right]. \qquad ...(1)$$

If n > 1, then 1 − n is negative and so n − 1 is positive.

Therefore in this case $\lim_{X \to \infty} X^{1-n} = \lim_{X \to \infty} \frac{1}{X^{n-1}} = \frac{1}{\infty} = 0.$

Hence from (1), we have

$$\int_a^{\infty} \frac{dx}{x^n} = \frac{a^{1-n}}{n-1}, \text{ if } n > 1.$$

Hence the given integral is convergent when n > 1.

If n < 1, then 1 − n is positive

and so $\lim_{X \to \infty} X^{1-n} = \infty.$

∴ Form (1), we have $\int_a^{\infty} \frac{dx}{x^n} = \infty.$

Hence the given integral is divergent when n < 1.

When n = 1, we have $\int_a^{\infty} \frac{dx}{x^n} = \int_a^{\infty} \frac{dx}{x}$

$$= \lim_{x \to \infty} \int_a^x \frac{dx}{x} = \lim_{x \to \infty} \left[\log x\right]_a^x = \lim_{x \to \infty} [\log x - \log a]$$

$$= \infty - \log a = \infty.$$

Hence the given integral is divergent when n = 1.

$\therefore \int_a^\infty \frac{dx}{x^n}$ converges when n > 1 and diverges when n ≤ 1.

ABEL'S TEST

If $\int_a^b f(x)dx$ *converges and* $\phi(x)$ *is bounded and monotonic for* $a \le x \le b$, *then* $\int_a^b f(x)\phi(x)dx$ *converges.*

ABLE'S TEST FOR THE CONVERGENCE OF INTEGRAL OF A PRODUCT

If $\int_a^\infty f(x)dx$ *converges and* $\phi(x)$ *is bounded and monotonic for* $x > a$, *then* $\int_a^\infty f(x)dx\, f(x)\, dx$ *is convergent.*

DIRICHLET'S TEST

If $\int_{a+\varepsilon}^b f(x)dx$ *be bounded and* $\phi(x)$ *be bounded and monotonic on the interval* $a \le x \le b$, *converging to zero as x tends to a, then* $\int_a^{b'} f(x)\phi(x)dx$ *converges.*

DIRICHLET'S TEST FOR THE CONVERGENCE OF INTEGRAL OF A PRODUCT

If $f(x)$ *be bounded and monotonic in the interval* $a \le x < \infty$ *and if* $\lim_{x \to \infty} f(x) = 0$, *then the integral* $\int_a^\infty f(x)\phi(x)dx$ *converges provided* $\left|\int_a^x \phi(x)dx\right|$ *is bounded as x takes all finite values.*

ABSOLUTE CONVERGENCE

The infinite integral $\int_a^\infty f(x)dx$ *is said to be absolutely convergent if the integral* $\int_a^\infty |f(x)|dx$ *is convergent.*

If the integral $\int_a^\infty f(x)dx$ is absolutely convergent, it is necessarily convergent. But if the integral $\int_a^\infty f(x)dx$ is convergent, it is not necessarily absolutely convergent. Thus, absolute convergence gives a sufficient but not a necessary condition for the convergence of an infinite integral.

COMPARISON TEST

Consider the improper integral $\int_a^b f(x)dx$, *where the range of integration (a, b) is finite and f (x) is unbounded only at x = a. Let g (x) be positive in the interval (a + ε, b) and | f (x) | ≤ g (x) in the interval (a + ε, b). Then* $\int_a^b f(x)dx$ *is convergent if* $\int_a^b g(x)dx$ *is convergent.*

Similarly if | f (x) | ≥ g (x) for all values of x in the interval (a + e, b), then $\int_a^b f(x)dx$ *is divergent provided* $\int_a^b g(x)dx$ *is divergent.*

Alternative Form of the Above Comparison Test

If $\lim_{x \to a} \frac{f(x)}{g(x)}$ is a definite number, other than zero, the integrals $\int_a^b f(x)dx$ and $\int_a^b g(x)dx$ either both converge or both diverge.

Note: While applying the above comparison test, we generally take $g(x) = \frac{1}{(x-a)^n}$ *i.e.,* $\int_a^b \frac{dx}{(x-a)^n}$ is generally taken as the comparison integral.

Theorem:

The comparison integral $\int_a^b \frac{dx}{(x-a)^n}$ *is convergent when n < 1 and divergent when n ≥ 1.*

Proof:

We have

$$\int_a^b \frac{dx}{(x-a)^n} = \lim_{\varepsilon \to 0} \int_{a+\varepsilon}^b \frac{dx}{(x-a)^n}$$

$$= \lim_{\varepsilon \to 0} \int_{a+\varepsilon}^b (x-a)^{-n} dx$$

$$= \lim_{\varepsilon \to 0} \left[\frac{(x-a)^{-n+1}}{1-n} \right]_{a+\varepsilon}^b \quad \text{if } n \neq 1$$

$$= \lim_{\varepsilon \to 0}\left[\frac{(b-a)^{1-n}}{1-n} - \frac{\varepsilon^{1-n}}{1-n}\right]. \qquad ...(1)$$

If $n < 1$, then $1 - n$ is positive and so $\lim_{\varepsilon \to 0} \varepsilon^{1-n} = 0$. Therefore from (1), we have

$$\int_a^b \frac{dx}{(x-a)^n} = \frac{(b-a)^{1-n}}{1-n}, \text{ if } n < 1.$$

Hence the given integral converges when $n < 1$.

If $n > 1$, then $1 - n$ is negative and so $n - 1$ is positive. Therefore in this case, from (1), we have

$$\int_a^b \frac{dx}{(x-a)^n} = \lim_{\varepsilon \to 0}\left[\frac{(b-a)^{1-n}}{1-n} + \frac{1}{(n-1)\varepsilon^{n-1}}\right] = \infty.$$

Hence the given integral diverges when $n > 1$.

when $n = 1$, we have

$$\int_a^b \frac{dx}{(x-a)^n} = \int_a^b \frac{dx}{(x-a)}$$

$$= \lim_{\varepsilon \to 0}\int_{a+\varepsilon}^b \frac{dx}{x-a} = \left[\log (x-a)\right]_{a+\varepsilon}^b$$

$$= \lim_{\varepsilon \to 0} [\log (b-a) - \log e]$$

$$= \infty, \qquad [\because \log 0 = -\infty]$$

Hence, the given integral diverges when $x = 1$.

TESTS FOR CONVERGENCE OF IMPROPER INTEGRALS OF THE SECOND KIND

Now we shall make a study of the tests for the convergence of a definite of the type $\int_a^b f(x)dx$ in which the range of integration is finite and the integrand $f(x)$ is unbounded at one or more points of the given interval [a, b]. It is sufficient to consider the case when $f(x)$ becomes abounded at $x = a$ and bounded for all other values of x in the interval [a, b]. In this case we have

$$\int_a^b f(x)dx = \lim_{\varepsilon \to 0}\int_{a+\varepsilon}^b f(x)dx.$$

In the articles to follow we give a few important tests for the convergence of the above integral.

ALTERNATIVE METHOD

Here $f(x) = \frac{1}{x\sqrt{(1+x^2)}} = \frac{1}{x^2\sqrt{\{1+(1/x^2)\}}}$. Take $g(x) = \frac{1}{x^2}$.

We have $\lim_{x\to\infty}\frac{f(x)}{g(x)} = \lim_{x\to\infty}\frac{1}{\sqrt{\{1+(1/x^2)\}}} = 1$, which is finite and non-zero therefore $\int_a^\infty f(x)dx$ and $\int_a^\infty g(x)dx$ either both converge or both diverge. But $\int_a^\infty g(x)dx = \int_a^\infty \frac{dx}{x^2}$ is convergent because here n = 2 which is > 1.

Hence $\int_a^\infty f(x)dx$ *i.e.*, $\int_a^\infty \frac{1}{x\sqrt{(1+x^2)}}dx$ is also convergent.

SOME DEFINITIONS

1. *Infinite Interval :* The interval whose length (range) is infinite is said to be an *infinite interval.* Thus, the intervals (a, ∞), (– ∞, b) and (– ∞, ∞) are infinite intervals.
2. *Bounded Functions :* A function f (x) is said to be *bounded* over the interval I if there exist two real numbers a and b (b > a) such that

$$a \le f(x) \le b \text{ for all } x \in I.$$

A function f (x) is said to be unbounded at a point, if it becomes infinite at that point. Thus the function

$$f(x) = x/\{(x-1)(x-2)\}$$

is unbounded each of the points x = 1 and x = 2.
3. *Monotonic Functions :* A real valued function f defined on an interval I is said to be *monotonically* increasing if

$$x > y \Rightarrow f(x) > f(x)\ \forall\ x, y \in I$$

and *monotonically decreasing* if

$$x > y \Rightarrow f(x) < f(x)\ \forall\ x, y \in I.$$

A function f defined on an interval I is said to be a monotonic function if it is either monotonically decreasing or monotonically increasing on I. For example the function f defined by f (x) = sin x is monotonically increasing in the interval $0 \le x \le 1/2\,\pi$ and monotonically decreasing in the interval $1/2\,\pi \le x \le \pi$.

4. *Proper Integral :* The definite integral $\int_a^b f(x)\,dx$ is said to be a *proper integral* if the range of integration is finite and the integrand f (x) is bounded. The integral $\int_0^{\pi/2} \sin x\,dx$ is a proper integral.

Also $\int_0^1 \frac{\sin x}{x} dx$ is an example of a proper integral because

$$\lim_{x \to 0} \frac{\sin x}{x} = 1.$$

5. *Improper Integral :* The definite integral $\int_a^b f(x)\,dx$ is said to be an *improper integral* if *(i)* the interval (a, b) is not finite (*i.e.*, is infinite) and the function f (x) is bounded over this interval; or *(ii)* the interval (a, b) is finite and f (x) is not bounded over this interval; or *(iii)* neither the interval (a, b) is finite nor f (x) is bounded over it.

6. *Improper Integrals of the First Kind or Infinite Integrals :* A definite integral $\int_a^b f(x)dx$ in which the range of integration is infinite (*i.e.*, either $b = \infty$ or $a = -\infty$ or both) and the integrand f (x) is bounded, is called an improper integral of the first kind or an infinite integral.

   Thus, $\int_0^\infty \frac{dx}{1+x^2}$ is an improper integral of the first kind since the upper limit of integration is infinite and the integrand $1/(1 + x^2)$ is bounded.

   Similarly, $\int_{-\infty}^0 e^x dx$ is an example of an improper integral of the first kind because here the lower limit of integration is infinite. Also $\int_{-\infty}^\infty \frac{dx}{1+x^2}$ is an improper integral of the first kind.

   In case the *interval (a, b) is infinite and the integrand f (x) is bounded,* we define

   (i) $\int_a^\infty f(x)dx = \lim_{x \to \infty} \int_a^x f(x)dx$, provided that the limit exists finitely *i.e.*, the limit is equal to a definite real number.

   (ii) $\int_{-\infty}^b f(x)dx = \lim_{x \to \infty} \int_{-x}^b f(x)dx$, provided that the limit exists finitely.

   (iii) $\int_{-\infty}^\infty f(x)dx = \lim_{x_1 \to \infty} \int_{-x_1}^c f(x)dx + \lim_{x_2 \to \infty} \int_c^{x_2} f(x)dx$

   provided that both these limits exist finitely.

7. *Improper Integrals of the Second Kind :* A definite integral $\int_a^b f(x)dx$ in which the range of integration is finite but the integrand f (x) is unbounded at one or more points of the interval $a \le x \le b$, is called an improper integral of the second kind.

Thus $\int_0^4 \frac{dx}{(x-2)(x-3)}$ and $\int_0^1 \frac{1}{x_2} dx$ are improper integrals of the second kind.

In the case of the definite integral

$$\int_a^b f(x)dx,$$

if the range of integration (a, b) is finite and the integrand $f(x)$ *is unbounded at one or more points of the given interval,* we define the value of the integral as follows:

(i) If f (x) is unbounded at x = b only *i.e.*, if f (x) → ∞ as x → b only, then we define

$\int_a^b f(x)dx = \lim_{\varepsilon \to 0} \int_a^{b-\varepsilon} f(x)dx$, provided that the limit exists finitely. Here ε is a small positive number.

(ii) If f (x) → ∞ as x → a only, then we define

$\int_a^b f(x)dx = \lim_{\varepsilon \to 0} \int_{a+\varepsilon}^{b} f(x)dx$, provided that the limit exists finitely.

(iii) If f (x) → ∞ as x → c only, where a < c < b, then we define

$\int_a^b f(x)dx = \lim_{\varepsilon \to 0} \int_a^{c-\varepsilon} f(x)dx + \lim_{\varepsilon \to 0} \int_{c+\varepsilon}^{b} f(x)dx$, provided that both these limits exist finitely.

(iv) If f (x) is unbounded at both the points and b of the interval (a, b) and is bounded at each other point of this interval, we write

$$\int_a^b f(x)dx = \int_a^c f(x)dx + \int_c^b f(x)dx,$$

where a < c < b and the value of the integral exists only if each of the integrals on the right hand side exists.

SOLVED EXAMPLES

Example 1:

Test the convergence of the integral $\int_1^2 \frac{dx}{\sqrt{(x^4-1)}}$.

Solution:

In the given integral the integrand $f(x) = 1/\sqrt{(x^4 - 1)}$ is anbounded at the lower limit of integration x = 1.

Take $g(x) = 1/\sqrt{(x^2 - 1)}$.

Then $\lim_{x \to 1} \frac{f(x)}{g(x)} = \lim_{x \to 1} \left\{ \frac{1}{\sqrt{(x^4-1)}} \cdot \sqrt{(x^2-1)} \right\}$

$$= \lim_{x \to 1} \frac{1}{\sqrt{(x^2+1)}}$$

$= 1/\sqrt{2}$, which is finite and non-zero.

Therefore by comparison test,

$$\int_1^2 f(x)dx \text{ and } \int_1^2 g(x)dx$$

are either both convergent or both divergent.

$$\text{But } \int_1^2 g(x)dx = \int_1^2 \frac{dx}{\sqrt{(x^2-1)}} = \lim_{\varepsilon \to 0} \int_{1+\varepsilon}^2 \frac{dx}{\sqrt{(x^2-1)}}$$

$$= \lim_{\varepsilon \to 0} \left[\log\{x + \sqrt{(x^2-1)}\}\right]_{1+\varepsilon}^2$$

$= \lim_{\varepsilon \to 0} [\log(2+\sqrt{3}) - \log\{1+\varepsilon+\sqrt{(\varepsilon^2+\varepsilon)}\}] = \log(2+\sqrt{3})$, which is a definite real number.

$\therefore \int_1^2 g(x)dx$ is convergent.

Hence $\int_1^2 \frac{1}{(x^4-1)} dx$ is also convergent.

Example 2:

Test the convergence of the integral

$$\int_0^1 \frac{dx}{(x+1)\sqrt{(1-x^2)}}.$$

Solution:

In the given integral the integrand $f(x) = \frac{1}{(x+1)\sqrt{(1-x^2)}}$ is unbounded at the upper limit of integration $x = 1$. Take

$$g(x) = \frac{1}{\sqrt{(1-x^2)}}$$

Then $\lim_{x \to 1} \frac{f(x)}{g(x)} = \lim_{x \to 1} \frac{1}{x+1} = \frac{1}{2}.$

Which is finite and non-zero.

Therefore, by comparison test.

$$\int_0^1 f(x)dx \text{ and } \int_0^1 g(x)dx$$

either both converge or both diverge.

But $\int_0^1 g(x)dx = \int_0^1 \frac{dx}{\sqrt{(1-x^2)}} = \lim_{\varepsilon \to 0} \int_0^{1-\varepsilon} \frac{dx}{\sqrt{(1-x^2)}}$

$= \lim_{\varepsilon \to 0} \left[\sin^{-1} x\right]_0^1 = \lim_{\varepsilon \to 0}$ [$\sin^{-1}$ (1 – ε) = $\sin^{-1} 1 = \pi/2$.

which is a definite real number.

$\therefore \int_0^1 g(x)dx$ is convergent.

Hence $\int_0^1 \frac{dx}{(x+1)\sqrt{(1-x^2)}}$ is also convergent.

Example 3:

Show that the integral $\int_0^1 \frac{\sec x}{x} dx$ *is divergent.*

Solution:

In the given integral the integrand $f(x) = \int_0^1 \frac{\sec x}{x} dx$ is unbounded at the lower limit of integration x = 0. Take g (x) = 1/x.

Then $\lim_{x \to 0} \frac{f(x)}{g(x)} = \lim_{x \to 0} \left\{\frac{\sec x}{x} \cdot x\right\} = \lim_{x \to 0} \sec x = 1,$

which is finite and non-zero.

Therefore, by comparison test,

$\int_0^1 f(x)dx$ and $\int_0^1 g(x)dx$

either both converge or both diverge. But the comparison integral $\int_0^1 \frac{1}{x} dx$ is divergent because here n = 1.

Hence the given integral $\int_0^1 \frac{\sec x}{x} dx$ is also divergent.

Example 4:

Discuss the convergence of the Beta function

$\int_0^1 x^{m-1}(1-x)^{n-1}\, dx.$

Solution:

Let $f(x) = x^{m-1}(1-x)^{n-1}$.

Example 5:

Show that the following integrals are convergent.

(i) $\int_0^\infty \frac{x^2}{(a^2+x^2)^2} dx$; (ii) $\int_1^\infty \frac{dx}{(1+x)\sqrt{x}}$.

Solution:

(i) Let b > 0. Then we have

$$\int_0^{\infty} \frac{x^2}{(a^2 + x^2)^2} dx = \int_0^{b} \frac{x^2}{(a^2 + x^2)^2} dx + \int_b^{\infty} \frac{x^2}{(a^2 + x^2)^2} dx.$$

The first integral on the right hand side is convergent because it is a proper integral. So we need to check the convergence of

$$\int_b^{\infty} \frac{x^2}{(a^2 + x^2)^2} dx.$$

Let $f(x) = \frac{x^2}{(a^2 + x^2)^2}$. Then f(x) is bounded in the integrals (b, ∞).

Take μ = 4 – 2 = 2. Then $\lim_{x \to \infty} x^{\mu} f(x) = \lim_{x \to \infty} x^2 \cdot \frac{x^2}{(a^2 + x^2)^2} =$

$\lim_{x \to \infty} \frac{1}{\{1 + (a^2 / x^2)\}^2} = 1,$

which is a definite real number.

Since μ > 1, therefore by μ-test $\int_b^{\infty} \frac{x^2}{(a^2 + x^2)^2} dx$ is convergent.

Hence $\int_0^{\infty} \frac{x^2 dx}{(a^2 + x^2)^2}$ is also convergent because it is the sum of two convergent integrals.

(ii) Apply μ-test by taking $\mu = \frac{3}{2}$.

Example 6:

Show that $\int_0^{\infty} \frac{dx}{(1 + x)^{2/3}}$ *is not convergent.*

Solution:

We have $\int_0^{\infty} \frac{dx}{(1 + x)^{2/3}}$

$= \lim_{x \to \infty} \int_0^{x} (1 + x)^{-2/3} dx$, (By def.)

$$= \lim_{x \to \infty} \left[\frac{(1 + x)^{1/3}}{\frac{1}{3}} \right]_0^x$$

$= \lim_{x \to \infty} 3\ [(1 + x)^{1/3} - 1] = \infty.$

Thus, the limit does not exist finitely and therefore the given integral is divergent (*i.e.*, the integral does not exist).

Example 7:

Test the convergence of

$$\text{(i)} \int_{-\infty}^{0} e^{x}dx; \qquad \text{(ii)} \quad \int_{-\infty}^{0} e^{-x}dx.$$

Solution:

(i) We have $\int_{-\infty}^{0} e^{x}dx = \lim_{x\to\infty} \int_{-x}^{0} e^{x}dx,$

$$= \lim_{x\to\infty} \left[e^{x}\right]_{-x}^{0} = \lim_{x\to\infty} [1 - e^{-x}] = [1 - 0] = 1.$$

Thus, the limits exist and is unique and finite; therefore the given integral is convergent.

(ii) We have $\int_{-\infty}^{0} e^{-x}dx = \lim_{x\to\infty} \int_{-x}^{0} e^{-x}dx,$

$$= \lim_{x\to\infty} \left[\frac{e^{-x}}{-1}\right]_{-x}^{0}$$

$$= -\lim_{x\to\infty} [e^{0} - e^{x}] = \infty.$$

Thus, the limit does not exist finitely and therefore the given integral is divergent (*i.e.*, the integral does not exist).

Example 8:

Test the convergence of the following integrals by evaluating them

$$\text{(i)} \int_{-\infty}^{0} \sinh x\, dx; \ \text{(ii)} \int_{-\infty}^{0} \cosh x\, dx.$$

Solution:

(i) We have $\int_{-\infty}^{0} \sinh x\, dx = \lim_{x\to\infty} \int_{-x}^{0} \sinh x\, dx$, (By def.)

$$= \lim_{x\to\infty} \int_{-x}^{0} \frac{e^{x} - e^{-x}}{2} dx$$

$$= \frac{1}{2}\left[\lim_{x\to\infty} \int_{-x}^{0} e^{x}dx - \lim_{x\to\infty} \int_{-x}^{0} e^{-x}dx\right]$$

$$= \frac{1}{2}\left[\lim_{x\to\infty} \left\{e^{x}\right\}_{-x}^{0} - \lim_{x\to\infty} \left\{\frac{e^{-x}}{-1}\right\}_{x}^{0}\right]$$

$$= \frac{1}{2}\left[\lim_{x \to \infty}\{e^0 - e^{-x}\} + \lim_{x \to \infty}\{e^0 - e^{x}\}\right] = \frac{1}{2}[1 - \infty] = -\infty.$$

Thus, the given integral diverges to $-\infty$.

(ii) We have

$$\int_{-\infty}^{0} \cosh x\, dx = \int_{-\infty}^{0} \frac{e^x + e^{-x}}{2} dx$$

$$= \frac{1}{2}\left[\int_{-\infty}^{0} e^x dx + \int_{-\infty}^{0} e^{-x} dx\right] = \frac{1}{2}[1 + \infty] = \infty.$$

Thus the given integral diverges to ∞.

Example 9:

Test the convergence of the integral $\int_1^\infty \frac{dx}{\sqrt{(x^3+1)}} dx$.

Solution:

Let $f(x) = \frac{1}{\sqrt{(x^3+1)}} = \frac{1}{x^{3/2}\sqrt{\{1+(1/x^2)\}}}$.

Take $g(x) = \frac{1}{x^{3/2}}$. We have $\lim_{x \to \infty} \frac{f(x)}{g(x)} = \lim_{x \to \infty} \frac{1}{\sqrt{\{1+(1/x^3)}} = 1,$

which is finite and non-zero. Therefore $\int_1^\infty f(x)dx$ and $\int_1^\infty g(x)dx$ are either both convergent or both divergent. But the comparison integral $\int_1^\infty g(x)dx$ *i.e.*, $\int_1^\infty \frac{dx}{x^{3/2}}$ is convergent because here $n = 3/2$ which is > 1.

Hence $\int_1^\infty f(x)dx$ *i.e.*, $\int_1^\infty \frac{dx}{\sqrt{(x^3+1)}}$ is also convergent.

Example 10:

Test the convergence of

(i) $\int_2^\infty \frac{dx}{\sqrt{(x^2 - x - 1)}}$; (ii) $\int_2^\infty \frac{dx}{\sqrt{(x^2-1)}}$.

Solution:

(i) Let $f(x) = \frac{1}{\sqrt{(x^2 - x - 1)}} = \frac{1}{\sqrt{\left\{\left(x - \frac{1}{2}\right)^2 - \frac{5}{4}\right\}}}$

Obviously f (x) = is bounded in the interval (2, ∞).

We can writ $f(x) = \frac{1}{x\sqrt{\{1-(1/x)-(1/x^2)\}}}$ Take $g(x) = \frac{1}{x}$.

We have

$$\lim_{x \to \infty} \frac{f(x)}{g(x)} = \lim_{x \to \infty} \frac{1}{\sqrt{\{1-(1/x)-(1/x^2)\}}} = 1,$$

which is finite and non-zero. Therefore

$$\int_2^\infty f(x)dx \text{ and } \int_2^\infty g(x)dx$$

either bot converge or both diverge.

But the comparison integral

$$\int_2^\infty g(x)dx \text{ i.e., } \int_2^\infty \frac{1}{x}dx$$

is divergent because here n = 1. Hence

$$\int_2^\infty f(x)dx \text{ i.e., } \int_2^\infty \frac{dx}{\sqrt{(x^2-x-1)}}$$

is also divergent.

(ii) Proceed exactly as inpart (i).

Here also the given integral is divergent.

Example 11:

Test the convergence of $\int_0^\infty \frac{x^3}{(x^2+a^2)^2}dx$.

Solution:

Let $f(x) = \frac{x^3}{(x^2+a^2)^2}dx$. Then f (x) is bounded in the interval (0, ∞).

Let b > 0.

We can write

$$\int_0^\infty \frac{d^2}{(x^2+a^2)^2}dx = \int_0^b \frac{x^3}{(x^2+a^2)^2}dx + \int_b^\infty \frac{x^3}{(x^2+a^2)}dx.$$

The first integral on the right hand side is a proper integral because the interval of integration (0, b) is finite and the integrand f (x) is bounded in this interval. So we need to check the convergence of the integral

$$\int_b^\infty \frac{x^3}{(x^2+a^2)^2}dx \text{ only.}$$

We can write

$$f(x) = \frac{x^3}{x^4\{1+(a^2/x^2)\}^2} = \frac{1}{x\{1+(a^2/x^2)\}^2}.$$

Take g (x) = 1/x.

We have $\lim_{x\to\infty}\frac{f(x)}{g(x)} = \lim_{x\to\infty}\frac{1}{\{1+(a^2/x^2)\}^2} = 1$, which is finite and non-zero. Therefore $\int_b^\infty f(x)dx$ and $\int_b^\infty g(x)dx$ either both converge or both diverge. But the comparison integral $\int_b^\infty g(x)dx$ *i.e.*, $\int_b^\infty \frac{1}{x}dx$ is divergent. Therefore, $\int_b^\infty f(x)dx$ is also divergent. Hence $\int_0^\infty \frac{x^3}{(x^2+a^2)^2}dx$ is divergent.

Example 12(a):

Test the convergence of the integral

$$\int_a^\infty \frac{sinx}{\sqrt{x}}dx, \text{ where } a > 0.$$

Solution:

Let $f(x) = \frac{1}{\sqrt{x}}$ and $\phi(x) = \sin x$.

Now $\frac{1}{\sqrt{x}}$ is bounded and monotonic decreasing for all $x \geq a$ and $\lim_{x\to\infty}\frac{1}{\sqrt{x}} = 0$.

Also $\left|\int_a^x \phi(x)dx\right| = \left|\int_a^x \sin x\, dr\right| = |\cos a - \cos x| \leq 2$, for all finite values of x. [Note that the value of cos x lies between – 1 and 1].

$\therefore \left|\int_a^x \phi(x)dx\right|$ is bounded for all finite values of x.

Hence by Dirichlet's test the integral $\int_0^\infty \frac{\sin x}{\sqrt{x}}dx$ is convergent.

Example 12(b):

Show that the integral $\int_a^\infty x^{n-1}e^{-x}dx$ *is convergent, where a > 0.*

Solution:

Do your self.

Example 12(c):

Test the convergence of the integral

$$\int_0^{\infty} \frac{x^{2m}}{1+x^{2n}}\,dx,$$

where m and n are positive integers.

Solution:

Let a > 0. We have

$$\int_0^{\infty} \frac{x^{2m}}{1+x^{2n}}\,dx = \int_0^{a} \frac{x^{2x}}{1+x^{2n}}\,dx + \int_a^{\infty} \frac{x^{2m}}{1+x^{2n}}\,dx.$$

The first integral on the right hand side is a proper integral and so it is convergent. Therefore the given integral is convergent or divergent according as $\int_a^{\infty} \frac{x^{2m}}{1+x^{2n}}\,dx$ is convergent or divergent.

To test the convergence of $\int_a^{\infty} \frac{x^{2m}}{1+x^{2n}}\,dx$,

let us take m = 2n – 2m.

We have $\lim_{x \to \infty} x^{\mu} \dfrac{x^{2m}}{1+x^{2n}} = \lim_{x \to \infty} x^{2n-2m} \dfrac{x^{2m}}{x^{2n}\{1+(1/x^{2n})\}}$

$$= \lim_{x \to \infty} \frac{1}{1+(1/x^{2n})} = 1, \text{ which is finite and non-zero.}$$

∴ by μ-test, the given integral is convergent if μ > 1 *i.e.*, if 2n – 2m > 1 which is possible if n > m since m and n are positive integers. Also by μ-test, the given integral is divergent if μ ≤ 1 *i.e.*, if 2n – 2m ≤ 1 *i.e.*, if n ≤ m since n and m are positive integers.

Example 13:

Test the convergence of

$$\int_0^{\infty} e^{-x} \frac{\sin x}{x}\,dx.$$

Solution:

We can write

$$\int_0^{\infty} e^{-x} \frac{\sin x}{x}\,dx = \int_0^{1} e^{-x} \frac{\sin x}{x}\,dx + \int_1^{\infty} e^{-x} \frac{\sin x}{x}\,dx.$$

Since $\lim_{x \to \infty} e^{-x} \dfrac{\sin x}{x} = 1$, therefore the integrand $e^{-x} \dfrac{\sin x}{x}$ is bounded throughout the finite interval (0, 1). So $\int_0^1 e^{-x} \frac{\sin x}{x}$ is a proper integral and therefore it is convergent. Thus we need to check the convergence of $\int_1^\infty e^{-x} \frac{\sin x}{x} dx$ only.

Let $f(x) = e^{-x} \frac{\sin x}{x}$. Then f (x) is bounded in the interval (1, ∞).

Take $g(x) = e^{-x}$. Then g (x) is positive in the interval (1, ∞).

We have

$$|f(x)| = \left|e^{-x} \frac{\sin x}{x}\right| = e^{-x}. |\sin x|. \frac{1}{2}$$

$$\leq e^{-x}, \text{ since } |\sin x| \leq 1 \text{ and } \frac{1}{x} \leq 1.$$

Thus | f (x) | g (x) throughout the interval (1, ∞).

∞ by comparison test $\int_1^\infty f(x)dx$ is convergent if $\int_1^\infty g(x)dx$ is convergent.

$$\text{Now } \int_1^\infty g(x)dx = \int_1^\infty e^{-x}dx = \lim_{x \to \infty} \int_1^x e^{-x}dx$$

$$= \lim_{x \to \infty} \left[-e^{-x}\right]_1^x$$

$$= \lim_{x \to \infty} [-e^{-x} + e^{-1}] = 0 + e^{-1} = 1/e,$$

which is a definite finite number. Hence $\int_1^\infty g(x)dx$ is convergent.

$\therefore \int_1^\infty f(x)dx$ is also convergent.

Hence $\int_0^\infty e^{-x} \frac{\sin x}{x} dx$ is convergent because the sum of two convergent integrals is also convergent.

Example 14:

Test the convergence of the integral

$$\int_0^\infty \frac{1 - \cos x}{x^2} dx.$$

Solution:

We have

$$\int_0^\infty \frac{1 - \cos x}{x^2} dx = \int_0^\infty \frac{2\sin^2 (x/2)}{x^2} dx$$

$$= \int_0^\infty \frac{2\sin^2 t}{4t^2} . 2dt,$$

putting $\frac{x}{2} = t$

so that $dx = 2dt$

$$= \int_0^\infty \frac{\sin^2 t}{t^2} dt.$$

Example 15(a):

Show that the integral $\int_0^\infty e^{-x^2} dx$ *is convergent.*

Solution:

We have

$$\int_0^\infty e^{-x^2} dx = \int_0^1 e^{-x^2} dx + \int_1^\infty e^{-x^2} dx.$$

Obviously $\int_0^1 e^{-x^2} dx$ is a proper integral because here the interval of integration (0, 1) is finite and the integrand e^{x^2} is bounded throughout this interval. Therefore this integral is convergent. So we need to check the convergence of $\int_1^\infty e^{-x^2} dx$ only.

Let $f(x) = e^{-x^2}$. Take $g(x) = xe^{-x^2}$ so that g (x) is positive throughout the interval $(1, \infty)$. We have

$$|f(x)| = e^{-x^2} \leq xe^{-x^2}, \text{ since } x \geq 1.$$

Thus $|f(x)| \leq g(x)$ throughout the interval $(1, \infty)$.

$\therefore$ by comparison test $\int_1^\infty e^{-x^2} dx$ is convergent if $\int_1^\infty xe^{-x^2} dx$ is convergent.

Now $\int_1^\infty xe^{-x^2} dx = \lim_{x \to \infty} \int_1^x xe^{-x^2} dx$

$$= \lim_{x \to \infty} \left(-\frac{1}{2}e^{-x^2}\right)_1^x = \lim_{x \to \infty} \left(-\frac{1}{2}e^{-x^2} + \frac{1}{2}e^{-1}\right)$$

$= \frac{1}{2}e^{-1}$, which is a definite number.

$\therefore \int_1^\infty xe^{-x^2} dx$ is convergent and so $\int_1^\infty e^{-x^2} dx$ is also convergent

Hence the given integral $\int_0^\infty e^{-x^2} dx$ is also convergent as it is the sum of two convergent integrals.

Example 15(b):

Examine the convergence of $\int_1^\infty \frac{dx}{x^{1/3}(1+x^{1/2})}$.

Solution:

Let $f(x) = \int_1^\infty \frac{dx}{x^{1/3}(1+x^{1/2})} = \frac{1}{x^{1/3}x^{1/2}\{1+(1/x^{1/2})\}}$

$= \frac{1}{x^{5/6}\{1+(1/x^{1/2})\}}$.

Obviously f(x) is bounded in the integral (1, ∞).

Take $\mu = \frac{5}{6} - 0 = \frac{5}{6}$. We have

$$\lim_{x\to\infty} x^{\mu} f(x) = \lim_{x\to\infty} x^{5/6} \cdot \frac{1}{x^{5/6}\{1+(1/x^{1/2})\}}$$

$$= \lim_{x\to\infty} \frac{1}{1+(1/x^{1/2})} = 1,$$

which is finite and non-zero. Since $\mu = \frac{5}{6}$ *i.e.*, < 1, it follows from the μ-test that the given integral is divergent.

Example 16:

Test the convergence of the integral

$$\int_0^\infty \frac{\cos mx}{x^2+^2} dx.$$

Solution:

Here $f(x) = \frac{\cos mx}{x^2+a^2}$. Let $g(x) = \frac{1}{x^2+a^2}$.

Obviously g (x) is positive in the interval (0, ∞).

we have $|f(x)| = \left|\frac{\cos mx}{x^2+a^2}\right| = \frac{|\cos mx|}{x^2+a^2}$

$\le \frac{1}{x^2+a^2}$., since $|\cos mx| \le 1$.

Thus $|f(x)| \le g(x)$ when $x \ge 0$.

∴ by comparison test, $\int_0^\infty \frac{\cos mx}{x^2+a^2} dx$ is convergent if $\int_0^\infty \frac{dx}{x^2+a^2} dx$ is convergent.

But $\int_0^\infty \frac{dx}{x^2+a^2} dx = \lim_{x\to\infty} \int_0^x \frac{dx}{x^2+s^2} = \lim_{x\to\infty}\left[\frac{1}{a}\tan^{-1}\frac{x}{a}\right]_0^x$

$$= \lim_{x \to \infty}\left[\frac{1}{a}\tan^{-1}\frac{x}{a} - 0\right] = \frac{1}{a}\cdot\frac{\pi}{2} = \text{a definite real number.}$$

$\therefore \int_0^{\infty} \frac{dx}{x^2 + a^2}$ is convergent.

Hence $\int_0^{\infty} \frac{\cos mx}{x^2 + a^2} dx$ is also convergent.

Example 17:

Test the convergence of the integral

$$\int_0^{\infty} \frac{\cos x}{1 + x^2} dx.$$

Solution:

Do your self.

Taking m = 1 and a = 1.

Example 18:

Test the convergence of the integral

$$\int_0^{\infty} \frac{\sin^2 x}{x^2} dx$$

Solution:

Let a > 0. Then we can write

$$\int_0^{\infty} \frac{\sin^2 x}{x^2} dx = \int_0^{a} \frac{\sin^2 x}{x^2} dx + \int_a^{\infty} \frac{\sin^2 x}{x^2} dx.$$

Since $\lim_{x \to 0} \frac{\sin^2 x}{x^2} = 1$, therefore the integrand $\frac{\sin^2 x}{x^2}$ is bounded throughout the finite interval (0, a).

So $\int_0^{a} \frac{\sin^2 x}{x^2} dx$ is a proper integral and we need to check the convergence of the integral $\int_0^{\infty} \frac{\sin^2 x}{x^2} dx$ only.

Here $f(x) = \frac{\sin^2 x}{x^2}$. Take $g(x) = \frac{1}{x^2}$.

Obviously g (x) is positive in the interval (a, ∞).

We have $|f(x)| = \left|\frac{\sin^2 x}{x^2}\right| = \frac{\sin^2 x}{x^2}$

$$\le \frac{1}{x^2}, \text{ since } \sin^2 x \le 1.$$

$\therefore$ by comparison test, $\int_a^\infty \frac{\sin^2 x}{x^2} dx$ is convergent is $\int_a^\infty \frac{dx}{x^2} dx$ is convergent.

But the comparison integral $\int_a^\infty \frac{dx}{x^2}$ is convergent because here n = 2 which is > 1.

$\therefore \int_a^\infty \frac{\sin^2 x}{x^2} dx$ is convergent.

Hence $\int_0^\infty \frac{\sin^2 x}{x^2} dx$ is convergent.

Example 19:

Show that the integral $\int_\pi^\infty \frac{\sin^2 x}{x^2} dx$ *is convergent.*

Solution:

Do your self.

By taking a = π.

Example 20(a):

Examine the convergence of $\int_0^\infty \frac{x\,dx}{(1+x)^3}$.

Solution:

Let a > 0. Then we have

$$\int_0^\infty \frac{x\,dx}{(1+x)^3} = \int_0^a \frac{x\,dx}{(1+x)^3} + \int_a^\infty \frac{x\,dx}{(1+x)^3}.$$

The first integral on the right hand side is convergent because it is a proper integral. We observe that in this in this integral the range of integrating (0, a) is finite and the integrand $x/(1+x)^3$ is bounded throughout the interval (0, a). So we need to check the convergence of $\int_a^\infty \frac{x\,dx}{(1+x)^3}$ only.

Let $f(x) = \frac{x}{(1+x)^3}$. Then f(x) is bounded in the interval (a, ∞).Take μ = 3 − 1 = 2. Then

$$\lim_{x \to \infty} x^\mu f(x) = \lim_{x \to \infty} x^2 \cdot \frac{x}{(1+x)^3}$$

$$= \lim_{x \to \infty} \frac{1}{\{1+(1/x)\}^3} = 1,$$

which exists *i.e.*, is equal to a definite real number.
Since $\mu = 2$ *i.e.*, > 1,

therefore by μ-test the integral $\int_a^{\infty} \frac{x\,dx}{(1+x)^3}$ is convergent.

Hence $\int_0^{\infty} \frac{x\,dx}{(1+x)^3}$ is also convergent because it is the sum of two convergent integrals.

Example 20(b):

Test the convergence of $\int_b^{\infty} \frac{x^{3/2}}{\sqrt{(x^4 - a^4)}}$, *where* $b > a$.

Solution:

Let $f(x) = \frac{x^{3/2}}{\sqrt{(x^4 - a^4)}}$. Then f (x) is bounded in the interval (b. ∞). Take

$$\mu = 2 - \frac{3}{2} = \frac{1}{2}. \text{ Then } \lim_{x \to \infty} x^{\mu} f(x) = \lim_{x \to \infty} x^{1/2} \cdot \frac{x^{3/2}}{x^2 \sqrt{\{1-(a^4 x^4)\}}}$$

$$= \lim_{x \to \infty} \frac{1}{\sqrt{\{1-(a^4 / x^4)\}}} = 1,$$

which is finite and non-zero.

Since $\mu < 1$, therefore by μ-test the given integral is divergent.

Example 20(c):

Examine the convergence of $\int_a^{\infty} \frac{dx}{x(\log x)^{n+1}}$, *where* $a > 1$.

Solution:

Let $\log x = t$ so that $(1/x)\,dx = dt$.

$$\therefore \int_a^{\infty} \frac{dx}{x(\log x)^{n+1}} = \int_{\log a}^{\infty} \frac{dt}{t^{n+1}}.$$

Let $f(t) = 1/t^{n+1}$. Then f (t) is bounded in the interval (log a, ∞).

Take $\mu = (n + 1) - 0 = n + 1$. Then

$$\lim_{t \to \infty} t^{\mu} f(t) = \lim_{t \to \infty} \frac{t^{n+1}}{t^{n+1}} = \lim_{t \to \infty} 1 = 1,$$

which is finite and non-zero.

Therefore by μ-test, the given integral is convergent if

$\mu > 1$ *i.e.*, $n + 1 > 1$ *i.e.*, $n > 0$

and divergent if $\mu \leq 1$ *i.e.*, $n + 1 \leq 1$ *i.e.*, $n \leq 0$.

Example 20(d):

Show that the integral $\int_1^\infty x^{n-1}e^{-x}dx$ *is convergent.*

Solution:

Let $f(x) = x^{n-1} e^{-x}$. Then f (x) is bounded in the interval $(1, \infty)$. We have

$$\lim_{x \to \infty} x^\mu f(x) = \lim_{x \to \infty} \frac{x^\mu . x^{n-1}}{e^x} = \lim_{x \to \infty} \frac{x^{\mu+n-1}}{1+x+\frac{x^2}{2!}+\ldots} = 0 \text{ for all values}$$

of μ and n.

Taking $\mu > 1$, we see by μ-test that the integral $\int_1^\infty x^{n-1}e^{-x}dx$

is convergent for all values of n.

Example 21:

Test the convergence of

$$\int_a^\infty (1-e^{-x})\frac{\cos x}{x^2}dx, \text{ when } a > 0.$$

Solution:

Let $f(x) = \frac{\cos x}{x^2}$ and $\phi(x) = 1 - e^{-x}$.

We have $\left|\frac{\cos x}{x^2}\right| \leq \frac{1}{x^2}$ as $|\cos x| \leq 1$.

Since $\int_a^\infty \frac{1}{x^2}dx$ is convergent, therefore by comparison test $\int_a^\infty \frac{\cos x}{x^2}dx$ is also convergent.

Again $\phi(x) = 1 - e^{-x}$ is monotonic increasing and bounded function for $x > a$.

Hence by Able's test $\int_a^\infty (1-e^{-x})\frac{\cos x}{x^2}dx$ is convergent.

Example 22:

Test the convergence of $\int_a^\infty e^{-x}\frac{\sin x}{x^2}dx$ *where* $a > 0$.

Solution:

Let $f(x) = \frac{\sin x}{x^2}$ and $\phi(x) = e^{-x}$.

Since $\left|\frac{\sin x}{x^2}\right| \le \frac{1}{x^2}$ and $\int_a^\infty \frac{1}{x^2}dx$ is convergent.

Again e^{-x} is monotonic decreasing and bounded function for x > a.

Hence by Able's test $\int_a^\infty e^{-x}\frac{\sin x}{x^2}dx$ is convergent, therefore by comparison test $\int_a^\infty \frac{\sin x}{x^2}dx$ is also convergent.

Example 23:

Show that $\int_a^\infty \sin x^2 dx$ *is convergent.*

Solution:

We have $\int_0^\infty \sin x^2 dx = \int_0^1 \sin x^2 dx + \int_1^\infty \sin x^2 dx$.

But $\int_0^1 \sin x^2 dx$ is a proper integral and hence convergent.

Now it remains to test the convergence of $\int_1^\infty \sin x^2 dx$. We can write

$$\int_1^\infty \sin x^2 dx = \int_1^\infty 2x.(\sin x^2).\frac{1}{2x}dx.$$

Let $f(x) = \frac{1}{2x}$ and $\phi(x) = 2x \sin x^2$.

The function $f(x) = \frac{1}{2x}$ is bounded and monotonic decreasing for all $x \ge 1$ and $\lim_{x \to \infty}\frac{1}{2x} = 0$.

Also $\left|\int_1^x \phi(x)dx\right| = \int_1^x 2x \sin x^2 dx$

$= |\cos 1^2 - \cos x^2| \le 2$, for all finite values of x.

$\therefore \left|\int_1^x \phi(x)dx\right|$ is bounded for all finite values of x.

Hence by Dirichlet's test

$$\int_1^\infty \frac{1}{2x}.(\sin x^2)2x\,dx$$

i.e., $\int_1^\infty \sin x^2 dx$ is convergent.

Since the sum of two convergent. Integrals is convergent, therefore the integral $\int_0^\infty \sin x^2 dx$ is convergent.

Example 24:

Show that the integral $\int_0^\infty \frac{\sin x}{x} dx$ *is convergent.*

Solution:

We have $\int_0^\infty \frac{\sin x}{x} dx = \int_0^a \frac{\sin x}{x} dx + \int_a^\infty \frac{\sin x}{x} dx$, where a > 0.

Since $\lim_{x \to \infty} \frac{\sin x}{x} = 1$, the integral $\int_0^a \frac{\sin x}{x} dx$ is a proper integral and hence convergent.

Now to test the convergence of $\int_a^\infty \frac{\sin x}{x} dx$.

Let $f(x) = 1/x$ and $\phi(x) = \sin x$.

The function $f(x) = 1/x$ is bounded and monotonic decreasing for all x $\geq a$ and $\lim_{x \to \infty} \frac{1}{x} = 0$.

Also $\left|\int_a^x \phi(x)\,dx\right| = \left|\int_a^x \sin x\,dx\right| = |\cos a - \cos x| \leq 2$, for all finite values of x.

$\therefore \left|\int_a^x \phi(x)\,dx\right|$ is bounded for all finite values of x.

Hence by Dirichlet's test the integral $\int_a^\infty \frac{\sin x}{x} dx$ is convergent.

Since the sum of two convergent integrals is convergent, therefore $\int_0^\infty \frac{\sin x}{x} dx$ is convergent.

Example 25:

Prove that $\int_a^\infty \frac{\cos \alpha x - \cos \beta x}{x} dx$ *is convergent where a > 0.*

Solution:

We have

$$\int_a^\infty \frac{\cos \alpha x - \cos \beta x}{x} dx$$

$$= \int_a^\infty \frac{\cos \alpha x}{x} dx - \int_a^\infty \frac{\cos \beta x}{x} dx.$$

The fiction $f(x) = 1/x$ is bounded and monotonic decreasing for all x $\geq a$ and $\lim_{x \to \infty} \frac{1}{x} = 0$.

Also $\left|\int_a^x \cos\alpha x\,dx\right| = \left|\frac{1}{\alpha}(\sin\alpha x - \sin\alpha a\right| \le \frac{2}{|\alpha|}$.

$\therefore \left|\int_a^x \cos\beta x\,dx\right|$ is bounded for all finite values of x.

Similarly $\left|\int_a^x \cos\beta x\,dx\right|$ is bounded for all finite values of x.

$\therefore$ by Dirichlet's test both the integrals

$$\int_a^\infty \frac{\cos\alpha x}{x}dx \text{ and } \int_a^\infty \frac{\cos\beta x}{x}dx \text{ are convergent.}$$

Hence the given integral is convergent.

Example 26:

Show that the integral

$$\int_a^\infty e^{-\alpha x}\frac{\sin x}{x}dx,\ a \ge 0 \text{ is convergent.}$$

Solution:

We have

$$\int_a^\infty e^{-\alpha x}\frac{\sin x}{x}dx = \int_0^\alpha e^{-\alpha x}\frac{\sin x}{x}dx + \int_\alpha^\infty e^{-\alpha x}\frac{\sin x}{x}dx,$$

where $\alpha > 0$.

Since $\lim_{x\to 0} e^{-\alpha x}\frac{\sin x}{x} = 1$, the integral $\int_0^\alpha e^{-\alpha x}\frac{\sin x}{x}dx$ is a proper integral and hence convergent.

Now it remains to test the convergence of

$$\int_\alpha^\infty e^{-\alpha x}\frac{\sin x}{x}dx.$$

Let $f(x) = \frac{e^{-ax}}{x}$ and $f(x) = \sin x$.

Obviously the function $f(x) = \frac{1}{xe^{ax}}$ is bounded and monotonic decreasing for all $x \ge \alpha$ and $\lim_{x\to\infty} f(x)$

$$= \lim_{x\to\infty}\frac{1}{xe^{ax}} = 0.$$

Moreover $\left|\int_\alpha^x \phi(x)dx\right| = \left|\int_\alpha^x \sin x\,dx\right| = |\cos a - \cos x| \le 2$, for all finite values of x.

$\therefore \left|\int_{\alpha}^{x} \phi(x)dx\right|$ is bounded for all finite values of x.

$\therefore$ by Dirichlet's test $\int_{\alpha}^{\infty} e^{-ax} \frac{\sin x}{x} dx$ is convergent.

Since the sum of two convergent integrals is convergent, therefore $\int_{0}^{\infty} e^{-ax} \frac{\sin x}{x} dx$ is convergent.

Example 27:

Show that $\int_{1}^{\infty} \frac{\sin x}{x^4} dx$ *is absolutely convergent.*

Solution:

The integral $\int_{1}^{\infty} \frac{\sin x}{x^4} dx$ will be absolutely convergent if $\int_{1}^{\infty} \left|\frac{\sin x}{x^4}\right| dx$ is convergent.

Let $f(x) = \left|\frac{\sin x}{x^4}\right|$. Then $f(x)$ is bounded in the interval $(1, \infty)$. We have

$$f(x) = \left|\frac{\sin x}{x^4}\right| = \frac{|\sin x|}{x^4} \leq \frac{1}{x^4}, \text{ since } |\sin x| \leq 1.$$

$\therefore$ by comparison test, $\int_{1}^{\infty} f(x)dx$ is convergent if $\int_{1}^{\infty} \frac{1}{x^4} dx$ is convergent.

But the comparison integral $\int_{1}^{\infty} \frac{1}{x^4} dx$ is convergent because here n = 4 which is > 1.

Hence $\int_{1}^{\infty} f(x)dx$ is convergent and so the given integral is absolutely convergent.

Example 28:

Show that $\int_{0}^{\infty} \frac{\sin mx}{a^2 + x^2} dx$ *converges absolutely.*

Solution:

The integral $\int_{0}^{\infty} \frac{\sin mx}{a^2 + x^2} dx$ will be absolutely convergent if $\int_{0}^{\infty} \left|\frac{\sin mx}{a^2 + x^2}\right| dx$ is convergent.

Let $f(x) = \left|\dfrac{\sin mx}{a^2 + x^2}\right|$. Then f(x) is bounded in the interval $(0, \infty)$.

We have

$$f(x) = \left|\frac{\sin mx}{a^2 + x^2}\right| = \frac{|\sin mx|}{a^2 + x^2} \le \frac{1}{a^2 + x^2}, \text{ since } |\sin mx| \le 1.$$

$\therefore$ by comparison test, $\int_0^\infty f(x)dx$ is convergent if $\int_0^\infty \frac{1}{a^2 + x^2} dx$ is convergent.

$$\text{But } \int_0^\infty \frac{dx}{a^2 + x^2} = \lim_{X \to \infty} \int_0^X \frac{dx}{a^2 + x^2}$$

$$= \lim_{X \to \infty} \left[\frac{1}{a}\tan^{-1}\frac{x}{a}\right]_0^X$$

$$= \lim_{X \to \infty} \left[\frac{1}{a}\tan^{-1}\frac{X}{a} - 0\right] = \frac{1}{a}\cdot\frac{\pi}{2},$$

which is a definite real number.

$\therefore \int_0^\infty \frac{dx}{a^2 + x^2}$ is convergent. Hence $\int_0^\infty f(x)dx$ is also convergent and so the given integral is absolutely convergent.

Example 29:

Show that the integral $\int_0^\infty e^{-x} \cos mx \, dx$ *converges absolutely.*

Solution:

The integral $\int_0^\infty e^{-x} \cos mx \, dx$ will be absolutely convergent if $\int_0^\infty \left|e^{-x} \cos mx\right| dx$ is convergent.

Let $f(x) = |e^{-x} \cos mx|$. Then f(x) is bounded in the interval $(0, \infty)$.

We have

$$f(x) = |e^{-x} \cos mx| = e^{-x} |\cos mx| \le e^{-x}, \text{ since } |\cos mx| \le 1.$$

$\therefore$ by comparison test, $\int_0^\infty f(x)dx$ is convergent if $\int_0^\infty e^{-x}dx$ is convergent.

$$\text{But } \int_0^\infty e^{-x}dx = \lim_{X \to \infty} \int_0^X e^{-x}dx = \lim_{X \to \infty}\left[-e^{-x}\right]_0^\infty$$

$= \lim_{X \to \infty} [-e^{-X} + 1] = 1$, which is a definite real number.

$\therefore \int_0^\infty e^{-x}dx$ is convergent.

Hence $\int_0^{\infty} f(x)dx$ is convergent and so the given integral is absolutely convergent.

Example 30:

Discuss the convergence of the following integrals by evaluating them

(i) $\int_1^{\infty} \frac{dx}{\sqrt{x}}$; (ii) $\int_1^{\infty} \frac{dx}{x^{3/2}}$.

Solution:

(i) We have

$$\int_1^{\infty} \frac{dx}{\sqrt{x}} = \lim_{x \to \infty} \int_1^{x} \frac{dx}{\sqrt{x}}, \text{ (By def.)}$$

$$= \lim_{x \to \infty} \int_1^{x} x^{-1/2} dx = \lim_{x \to \infty} \left[\frac{x^{1/2}}{1/2}\right]_1^x$$

$$= \lim_{x \to \infty} [2\sqrt{x} - 2] = \infty$$

Thus, the limit does not exist finitely and therefore the given integral is divergent (*i.e.*, the integral does not exist).

(ii) We have

$$\int_0^{\infty} \frac{dx}{x^{3/2}} = \lim_{x \to \infty} \int_1^{x} \frac{dx}{x^{3/2}},$$

$$= \lim_{x \to \infty} \int_1^{x} x^{-3/2} dx = \lim_{x \to \infty} \left[\frac{x^{-1/2}}{-1/2}\right]_1^x$$

$$= \lim_{x \to \infty} \left[-\frac{2}{\sqrt{x}}\right]_1^x$$

$$= \lim_{x \to \infty} \left[-\frac{2}{\sqrt{x}} + 2\right] = 2.$$

Thus, the limit exists and is unique and finite; therefore the given integral is convergent and its value is 2.

Example 31(a):

Evaluate $\int_1^{\infty} \frac{dx}{x}$.

Solution:

We have

$$\int_1^{\infty} \frac{dx}{x} = \lim_{x \to \infty} \int_1^{x} \frac{dx}{x} = \lim_{x \to \infty} [\log x]_1^x$$

$$= \lim_{x \to \infty}[\log x - 0] = \infty.$$

Thus, the limit does not exist finitely and therefore the given integral is divergent (*i.e.*, the integral does not exist).

Example 31(b):

Evaluate $\int_3^{\infty} \frac{dz}{(x-2)^2}$.

Solution:

We have

$$\int_3^{\infty} \frac{dz}{(x-2)^2} = \lim_{x \to \infty} \int_3^{x} \frac{dx}{(x-2)^2},$$

$$= \lim_{x \to \infty} \int_3^{x} (x-2)^{-2}\,dx = \lim_{x \to \infty}\left[\frac{(x-2)^{-1}}{-1}\right]_3^x$$

$$= \lim_{x \to \infty}\left[-\frac{1}{x-2}\right]_3^x = \lim_{x \to \infty}\left[-\frac{1}{x-2}+1\right] = 1,$$

Which is a definite real number. Therefore the given integral is convergent and its value is 1.

Example 32:

Test the convergence of $\int_0^{\infty} e^{-mx}dx,$ (m > 0).

Solution:

We have $\int_0^{\infty} e^{-mx}dx = \lim_{x \to \infty} \int_0^{x} e^{-mx}dx,$ (By def.)

$$= \lim_{x \to \infty}\left[\frac{e^{-mx}}{-m}\right]_0^x = \lim_{x \to \infty}\left\{-\frac{1}{m}(e^{-mx}-1\right\} = -\frac{1}{m}[0-1] = \frac{1}{m}.$$

Thus, the limit exists and is unique and finite, therefore the given integral is convergent.

Example 33:

Test the convergence of $\int_0^{\infty} e^{2x}dx,$

Solution:

We have $\int_0^{\infty} e^{2x}dx = \lim_{x \to \infty} \int_0^{x} e^{2x}dx,$ (By def.)

$$= \lim_{x \to \infty}\left[\frac{e^{2x}}{2}\right]_0^x = \frac{1}{2}\lim_{x \to \infty}[e^{2x}-1] = \infty.$$

Thus the limit does not exist finitely and therefore the given integral is divergent (*i.e.*, the integral does not exist).

Example 34:

Test the convergence of $\int_0^\infty \frac{4a\,dx}{x^2 + 4a^2}$.

Solution:

We have $\int_0^\infty \frac{4a\,dx}{x^2 + 4a^2} = \lim_{x \to \infty} \int_0^x \frac{4a\,dx}{x^2 + (2a)^2}$, (By def.)

$$= \lim_{x \to \infty} \left[4a \cdot \frac{1}{2a} \tan^{-1} \frac{x}{2a}\right]_0^x = 2 \lim_{x \to \infty} \left[\tan^{-1} \frac{x}{2a}\right]_0^x$$

$$= 2 \cdot \lim_{x \to \infty} \left[\tan^{-1} \frac{x}{2a} - 0\right] = 2.\ [\tan^{-1} \infty] = 2.\frac{\pi}{2} = \pi.$$

Thus, the limit exists and is unique and finite; therefore, the given integral is convergent.

Example 35:

Test the convergence of $\int_{-\infty}^{\infty} e^{-x} dx$.

Solution:

We have

$$\int_{-\infty}^{\infty} e^{-x} dx = \int_{-\infty}^{0} e^{-x} dx + \int_0^\infty e^{-x} dx$$

$$= + \infty + 1 = \infty.j$$

Thus, the limit does not exist finitely. Therefore the given integral is divergent (*i.e.*, the integral does not exist).

Example 36(a):

Test the convergence of $\int_{-\infty}^{\infty} \frac{dx}{1 + x^2}$

Solution:

We have

$$\int_{-\infty}^{\infty} \frac{dx}{1 + x^2} = \int_{-\infty}^{\infty} \frac{dx}{1 + x^2} + \int_0^\infty \frac{dx}{1 + x^2}$$

$$= \lim_{x \to \infty} \int_{-x}^{0} \frac{dx}{1 + x^2} + \lim_{x \to \infty} \int_0^x \frac{dx}{1 + x^2}$$

$$= \lim_{x \to \infty} \left[\tan^{-1} x\right]_{-x}^{0} + \lim_{x \to \infty} \left[\tan^{-1} x\right]_0^x$$

$$= \lim_{x \to \infty} [0 - \tan^{-1}(-x)] + \lim_{x \to \infty}[\tan^{-1} x - 0]$$

$$= -(-\pi/2) + \pi/2 = \pi.$$

Thus, the limit exists and is unique and finite; therefore the given integral is convergent.

Example 36(b):

Evaluate $\int_{-\infty}^{\infty} \frac{dx}{x^2 + 2x + 2}$.

Solution:

We have $\int_{-\infty}^{\infty} \frac{dx}{x^2 + 2x + 2} = \int_{-\infty}^{\infty} \frac{dx}{(x+1)^2 + 1}$

$$= \lim_{x_1 \to \infty} \int_{-x_1}^{c} \frac{dx}{(x+1)^2 + 1} + \lim_{x_2 \to \infty} \int_{c}^{x_2} \frac{dx}{(x+1)^2 + 1},$$

Where c is any real number

$$= \lim_{x_1 \to \infty} \left[\tan^{-1}(x+1)\right]_{-x_1}^{c} + \lim_{x_2 \to \infty} \left[\tan^{-1}(x+1)\right]_{c}^{x_2}$$

$$= \lim_{x_1 \to \infty} [\tan^{-1}(c+1) + \tan^{-1}(1 - x_1)]$$

$$+ \lim_{x_2 \to \infty} [\tan^{-1}(x_2 + 1) - \tan^{-1}(c+1)]$$

$$= \tan^{-1}(c+1) - \tan^{-1}(-\infty) + \tan^{-1}(\infty) - \tan^{-1}(c+1)$$

$$= -(-\pi/2) + \pi/2 = \pi.$$

Hence the given integral is convergent and its value is π.

Example 37:

Evaluate $\int_0^1 \frac{dx}{\sqrt{x}}$.

Solution:

In the given integral, the integrand $1/\sqrt{x}$ becomes infinite at the lower limit $x = 0$. Therefore we have

$$\int_0^1 \frac{dx}{\sqrt{x}} = \lim_{\varepsilon \to 0} \int_{0+\varepsilon}^{1} \frac{dx}{\sqrt{x}}$$

$$= \lim_{\varepsilon \to 0} \left[2\sqrt{x}\right]_{\varepsilon}^{1}$$

$$= \lim_{\varepsilon \to 0} [2 - 2\sqrt{\varepsilon}] = 2.$$

Hence the given integral is convergent and its value is 2.

Example 38(a):

Evaluate $\int_{-1}^{1} \frac{dx}{x^2}$.

Solution:

Here the integrand becomes infinite at x = 0 and – 1 < 0 < 1.

$$\therefore \int_{-1}^{1} \frac{dx}{x^2} = \lim_{\varepsilon \to 0} \int_{-1}^{-\varepsilon} \frac{dx}{x^2} + \lim_{\varepsilon' \to 0} \int_{\varepsilon'}^{1} \frac{dx}{x^2}$$

$$= \lim_{\varepsilon \to 0} \left[-\frac{1}{x} \right]_{-1}^{-\varepsilon} + \lim_{\varepsilon' \to 0} \left[-\frac{1}{x} \right]_{\varepsilon'}^{1}$$

$$= \lim_{\varepsilon \to 0} \left[\frac{1}{\varepsilon} - 1 \right] + \lim_{\varepsilon' \to 0} \left[-1 + \frac{1}{\varepsilon'} \right].$$

Since both the limits do not exist finitely, therefor the integral does note exist and is divergent.

Example 38(b):

Evaluate $\int_{0}^{2a} \frac{dx}{(x-a)^2}$.

Solution:

Here the integrand becomes infinite at x = a and 0 < a < 2a. We find that the limits do not exist finitely *i.e.*, the given integral is meaningless.

Example 39:

Shwo that the integral $\int_{0}^{1} \frac{dx}{x^{1/3}(1+x^2)}$ *is convergent.*

Solution:

In the given integral, the integrand $f(x) = \int_{0}^{1} \frac{dx}{x^{1/3}(1+x^2)}$ is unbounded at the lower limit of integration x = 0.

Take $g(x) = 1/x^{1/3}$.

Then $\lim_{x \to 0} \frac{f(x)}{g(x)} = \lim_{x \to 0} \frac{1}{1+x^2} = 1$, which is finite and non-zero.

∴ by comparison test

$$\int_{0}^{1} f(x)dx$$

and $\int_{0}^{1} g(x)dx$

either both converge or both diverge. But the comparison integral $\int_0^1 \frac{dx}{x^{1/3}}$ is convergent because here n = 1/3 which is less than is less than 1. Hence the integral $\int_0^1 \frac{dx}{x^{1/3}(1+x^2)}$ is also convergent.

Example 40:

Test the convergence of the integral

$$\int_\pi^\infty \frac{\sin x}{x^2}\,dx.$$

Solution:

Let $f(x) = \frac{\sin x}{x^2}$. Then g (x) is positive in the interval (π, ∞). Take (x) = $1/x^2$. Then g(x) is positive in the interval (p, ∞).

We have $|f(x)| = \left|\frac{\sin x}{x^2}\right| = \frac{|\sin x|}{x^2}$

$\leq \frac{1}{x^2}$, since $|\sin x| \leq 1$.

∴ by comparison test, $\int_\pi^\infty \frac{\sin x}{x^2}\,dx$ is convergent is $\int_\pi^\infty \frac{dx}{x^2}$ is convergent.

But the comparison integral $\int_\pi^\infty \frac{dx}{x^2}$ is convergent because here n = 2 which is > 1.

Hence $\int_\pi^\infty \frac{\sin x}{x^2}\,dx$ is also convergent.

Example 41:

Show that the integral $\int_a^\infty \frac{dx}{x\sqrt{(1+x^2)}}$ *converges, where a > 0.*

Solution:

Let $f(x) = \frac{1}{x\sqrt{(1+x^2)}}$.

Then f (x) is bounded in the interval (a, ∞). Take g (x) = $1/x^2$.

Then g (x) is positive in the interval (a, ∞). We have

$$|f(x)| = \left|\frac{1}{x\sqrt{(1+x^2)}}\right| = \frac{x}{x^2\sqrt{\{1+(1/x^2)\}}}$$

$< \frac{1}{x^2}$, since $\frac{1}{\sqrt{\{1+(1/x^2)\}}} < 1$.

$\therefore$ by the comparison integral $\int_a^\infty \frac{dx}{x\sqrt{(1+x^2)}}$ is convergent if $\int_a^\infty \frac{dx}{x^2}$ is convergent.

But comparison integral $\int_a^\infty \frac{dx}{x^2}$ is convergent because here n = 2 which is > 1.

Hence $\int_a^\infty \frac{dx}{x\sqrt{(i+x^2)}}$ is also convergent.

Example 42(a):

Test the convergent of $\int_0^2 \frac{\log x}{\sqrt{(2-x)}} dx$.

Solution:

Let $f(x) = \frac{\log x}{\sqrt{(2-x)}} dx$. Then f(x) is unbounded both at x = 0 and x = 2. If 0 < a < 2, we can write

$$\int_0^2 \frac{\log x}{\sqrt{(2-x)}} dx$$

$$= \int_0^a \frac{\log x}{\sqrt{(2-x)}} dx + \int_a^2 \frac{\log x}{\sqrt{(2-x)}} dx$$

$$= I_1 + I_2, \text{ say.}$$

To test the convergence of I_1. We have

$$\lim_{x \to 0} x^\mu f(x) = \lim_{x \to 0} \left\{ x^\mu \cdot \frac{\log x}{\sqrt{(x-2)}} \right\} = 0 \text{ if } \mu > 0.$$

Therefore taking μ between 0 and 1, it follows by μ-test that I_1 is convergent.

To test the convergence of I_2. Take $\mu = \frac{1}{2}$. We have

$$\lim_{x \to 2-0} (2-x)^\mu . f(x)$$

$$= \lim_{x \to 2-0} (2-x)^{1/2} \cdot \frac{\log x}{\sqrt{(2-x)}}$$

$$= \lim_{x \to 2-0} \log x$$

$$= \lim_{\varepsilon \to 0} \log(2-\varepsilon) = \log 2.$$

$\therefore$ by μ-test I_2 is convergent because 0 < μ < 1.

Hence, the given integral is also convergent, it being the sum of two convergent integrals.

Example 42(b):

Test the convergence of $\int_0^1 x^{p-1} e^{-x} dx$.

Solution:

Let $f(x) = x^{p-1} e^{-x}$ and $I = \int_0^1 x^{p-1} e^{-x} dx$.

If $p \geq 1$, $f(x)$ is bounded throughout the interval $(0, 1)$ and so I is a proper integral and hence it is convergent if $p \geq 1$.

If $p < 1$, $f(x)$ is unbounded at $x = 0$. In this case, we have

$$\lim_{x \to 0} x^{\mu} f(x) = \lim_{x \to 0} x^{\mu} . x^{p-1} e^{-x} = \lim_{x \to 0} x^{\mu+p-1} e^{-x}$$

$= 1$ is $\mu + p - 1 = 0$ *i.e.*, $\mu = 1 - p$.

So by μ-test when $0 < \mu < 1$ *i.e.*, $0 < p < 1$, the given integral is convergent and when $\mu \geq 1$ *i.e.*, $p \leq 0$, the given integral is divergent.

Hence I is convergent if $p > 0$ and is divergent if $p \leq 0$.

Example 43:

Test the convergence of the integral $\int_0^1 \frac{dx}{x^{1/3}(1+x^2)}$.

Solution:

In the given integral the integrand $f(x) = \frac{dx}{x^3(1+x^2)}$ is unbounded at the lower limit of integration $x = 0$. Take $g(x) = 1/x^3$.

Then $\lim_{x \to 0} \frac{f(x)}{g(x)}$

$$= \lim_{x \to 0} \frac{1}{1+x^2} = 1,$$

which is finite and non-zero. Therefore, by comparison test,

$$\int_0^1 f(x)dx \text{ and } \int_0^1 g(x)dx$$

either both converge or both diverge. But the comparison integral $\int_0^1 \frac{dx}{x^3}$ is divergent because here $n = 3$ which is > 1.

Hence the given integral $\int_0^1 \frac{dx}{x^3(1+x^2)}$ is also divergent.

Example 44:

Test the convergence of the integral $\int_0^{\pi/2} \frac{\cos x}{x^2} dx$.

Solution:

In the given integral the integrand $f(x) = \frac{\cos x}{x^2}$ is unbounded at the lower limit of integration $x = 0$. Take $g(x) = 1/x^2$.

Then $\lim_{x \to 0} \frac{f(x)}{g(x)} = \lim_{x \to 0} \left\{ \frac{\cos x}{x^2} \cdot x^2 \right\} = \lim_{x \to 0} \cos x = 1$,

which is finite and non-zero.

Therefore, by comparison test, $\int_0^{\pi/2} f(x)dx$ and $\int_0^{\pi/2} g(x)dx$ either both converge or both diverge.

$$\text{But } \int_0^{\pi/2} g(x)dx = \int_0^{\pi/2} \frac{1}{x^2} dx = \lim_{\varepsilon \to 0} \int_\varepsilon^{\pi/2} \frac{1}{x^2} dx$$

$$= \lim_{\varepsilon \to 0} \left[-\frac{1}{x} \right]_\varepsilon^{\pi/2} = \lim_{\varepsilon \to 0} \left[-\frac{2}{\pi} + \frac{1}{\varepsilon} \right] = \infty.$$

$\therefore \int_0^{\pi/2} g(x)dx$ is divergent.

Hence the given integral $\int_0^{\pi/2} \frac{\cos x}{x^2} dx$ is also divergent.

Example 45:

Show that $\int_0^1 x^{n-1} e^{-x} dx$ *is convergent if* $n > 0$.

Solution:

If $n \geq 1$, then $\int_0^1 x^{n-1} e^{-x} dx$ is a proper integral because the integrand $f(x) = x^{n-1} e^{-x}$ is bounded in the interval $(0, 1)$. So the given integral is convergent when $n \leq 1$.

If $0 < n < 1$, the integrand $f(x) = x^{n-1}e^{-x}$ is unbounded at $x = 0$.

Take $g(x) = x^{n-1}$.

Then $\lim_{x \to 0} \frac{f(x)}{g(x)} = \lim_{x \to 0} \frac{f(x)}{g(x)} e^{-x} = 1$, which is finite and non-zero.

$\therefore$ by comparison test, $\int_0^1 f(x)dx$ and $\int_0^1 g(x)dx$ either both converge or both diverge.

$$\text{But } \int_0^1 g(x)dx = \int_0^1 x^{n-1}dx = \lim_{\varepsilon \to 0} \int_\varepsilon^1 x^{n-1}dx = \lim_{\varepsilon \to 0} \left[\frac{x^n}{n} \right]_\varepsilon^1$$

$$= \lim_{\varepsilon \to 0}\left[\frac{1}{n} - \frac{\varepsilon^n}{n}\right] = \frac{1}{n}, \text{ which is a definite real number.}$$

$\therefore \quad \int_0^1 g(x)dx$ is convergent.

Hence $\int_0^1 x^{n-1}e^{-x}dx$ is also convergent.

Example 46:

Show that the integral $\int_0^\infty x^{n-1}e^{-x}dx$ *is convergent if* $n > 0$.

Solution:

We have $\int_0^\infty x^{n-1}e^{-x}dx = \int_0^1 x^{n-1}e^{-x}dx + \int_1^\infty x^{n-1}e^{-x}dx.$

Let $I_1 = \int_0^1 x^{n-1}e^{-x}dx$ and $I_2 = \int_1^\infty x^{n-1}e^{-x}dx.$

The integral I_2 is convergent for all vaues of n.

Also the integral I_1 is convergent if $n > 0$.

Hence the given integral is convergent if $n > 0$ because then it is the sum of two convergent integrals.

Example 47(a):

Prove that the integral $\int_0^1 \frac{dx}{\sqrt{\{x(1-x)\}}}$ *converges.*

Solution:

In the given integral the integrand $f(x) = 1/\sqrt{\{x(1-x)\}}$ is unbounded both at $x = 0$ and at $x = 1$. If $0 < a < 1$, we can write

$$\int_0^1 \frac{dx}{\sqrt{\{x(1-x)\}}} = \int_0^a \frac{dx}{\sqrt{\{x(1-x)\}}} + \int_a^1 \frac{dx}{\sqrt{\{x(1-x)\}}}$$

$= I_1 + I_2$, say.

In the integral I_1 the integrand $f(x)$ is unbounded at the lower limit of integration $x = 0$ and in the integral I_2 the integrand $f(x)$ is unbounded at the upper limit of integration $x = 1$.

To test the convergence of I_1. Take $\mu = \frac{1}{2}$. We have

$$\lim_{x \to 0} x^\mu f(x) = \lim_{x \to 0} x^{1/2} \cdot \frac{1}{\sqrt{\{x(1-x)\}}}$$

$$= \lim_{x \to 0} \frac{1}{\sqrt{(1-x)}}$$

$= 1$ *i.e.*, the limit exists.

Since $0 < \mu < \frac{1}{2}$, therefore by μ-test I_1 is convergent.

To test the convergence of I_2. Take $\mu = \frac{1}{2}$. We have

$$\lim_{x \to 1-0} (1-x)^{\mu} . f(x)$$

$$= \lim_{x \to 1-0} (1-x)^{1/2} \cdot \frac{1}{\sqrt{\{x(1-x)\}}}$$

$$= \lim_{x \to 1-0} \frac{1}{\sqrt{x}} = \lim_{\varepsilon \to 0} \frac{1}{\sqrt{(1-\varepsilon)}} = 1.$$

Hence by μ-test I_2 is convergent since $0 < \mu < 1$.

Thus the given integral is the sum of two converged integrals.

Hence the given integral itself is convergent.

Example 47(b):

Test the convergence of $\int_0^{\pi/4} \frac{1}{\sqrt{(tan x)}} dx$.

Solution:

Here the integrand $f(x) = 1 \sqrt{(\tan x)}$ is unbounded at $x = 0$.

Take $\mu = \frac{1}{2}$.

We have $\lim_{x \to 0} x^{\mu} . f(x)$

$$= \lim_{x \to 0} x^{1/2} \cdot \frac{1}{\sqrt{(\tan x)}}$$

$$= \lim_{x \to 0} \sqrt{\left(\frac{x}{\sin x}\right)} \cdot \sqrt{(\cos x)}$$

$= 1.1 = 1.$

Since $0 < \mu < 1$, therefore by μ-test the given integral is convergent.

Example 48:

Test the convergence of $\int_0^{\pi/2} \frac{\sin x}{x^{1+n}} dx$.

Solution:

We have $\int_0^{\pi/2} \frac{\sin x}{x^{1+n}} dx$

$$= \int_0^{\pi/2} \left(\frac{\sin x}{x}\right) \cdot \frac{1}{x^n} dx.$$

Now $\lim_{x \to 0} \frac{\sin x}{x} = 1$. Therefore, for $n \le 0$, the integrand is bounded throughout the interval $(0, \pi/2)$ and so the given integral is a proper integral and hence it is convergent if $n \le 0$.

If $n > 0$, the integrand is unbounded only at $x = 0$. In this case, we have

$$\lim_{x \to 0} x^\mu \frac{\sin x}{x^{n+1}}$$

$$= \lim_{x \to 0} \left\{\left(\frac{\sin x}{x}\right) \cdot \frac{x^\mu}{x^n}\right\}$$

$$= \lim_{x \to 0} \left\{x^{\mu - n} \cdot \left(\frac{\sin x}{x}\right)\right\} = 1,$$

if $\mu - n$ 0 *i.e.*, $\mu = n$.

$\therefore$ by μ-test if $0 < \mu < 1$ *i.e.*, $0 < n < 1$, the given integral is convergent and if $\mu \ge 1$ *i.e.*, $n \ge 1$, the given integral is divergent.

Hence the given integral is convergent if $n < 1$ and divergent if $n \ge 1$.

Example 49(a):

Test the convergence of $\int_0^{\pi/2} \frac{\cos x}{x^n} dx$.

Solution:

When $n \le 0$, the given integral is a proper integral and hence convergent.

When $n > 0$, the integrand becomes unbounded at $x = 0$.

Let $f(x) = \frac{\cos x}{x^n}$.

Then $\lim_{x \to 0} x^\mu f(x) = \lim_{x \to 0} x^{\mu - n} \cos x = 1$, if $\mu = n$.

Hence by μ-test it follows that the given integral is convergent when $0 < n < 1$, and divergent when $n \ge 1$.

From the above discussion we conclude that the given integral is convergnet when $n < 1$, and divergent when $n \ge 1$.

Example 49(b):

Show that the integral $\int_0^{\pi/2} \log \sin x \, dx$ *converges.*